Birth and Fortune

The author gratefully acknowledges permission to reprint from the following sources:

H. Scott Gordon, "On Being Demographically Lucky: The Optimum Time to Be Born," Presidential Address to the Western Economic Association, Annual Meeting, Anaheim, California (June 1977), p. 4; Appendix Table 6.1 from Joseph Veroff, "General Feelings of Well Being Over a Generation, 1957–1976," unpublished paper presented to American Psychological Association (1 September 1978); A. Regula Herzog, Jerald G. Bachman, and Lloyd D. Johnson, "High School Seniors Preferences for Sharing Work and Family Responsibilities between Husband and Wife" (Monitoring the Future Occasional Paper 3), Ann Arbor, Michigan: Institute for Social Research (1978), Appendix Table 3.

304.6
E13b

Library of Congress Cataloging in Publication Data

Easterlin, Richard Ainley, 1926–
    Birth and fortune.

    Includes bibliographical references and index.
    1. United States—Population.   2. Fertility, Human—
United States.   3. United States—Economic conditions—
1945–   4. United States—Social conditions—1945–
I. Title.
HB3505.E247   304.6   79–56369
ISBN: 0–465–00688–4

265734

Birth and Fortune

BIRTH AND FORTUNE

The Impact of Numbers on Personal Welfare

RICHARD A. EASTERLIN

Basic Books, Inc., Publishers   New York

The author gratefully acknowledges permission to reprint from the following sources:

H. Scott Gordon, "On Being Demographically Lucky: The Optimum Time to Be Born," Presidential Address to the Western Economic Association, Annual Meeting, Anaheim, California (June 1977), p. 4; Appendix Table 6.1 from Joseph Veroff, "General Feelings of Well Being Over a Generation, 1957–1976," unpublished paper presented to American Psychological Association (1 September 1978); A. Regula Herzog, Jerald G. Bachman, and Lloyd D. Johnson, "High School Seniors Preferences for Sharing Work and Family Responsibilities between Husband and Wife" (Monitoring the Future Occasional Paper 3), Ann Arbor, Michigan: Institute for Social Research (1978), Appendix Table 3.

Library of Congress Cataloging in Publication Data

Easterlin, Richard Ainley, 1926–
   Birth and fortune.

   Includes bibliographical references and index.
   1.   United States—Population.   2.   Fertility, Human—United States.   3.   United States—Economic conditions—1945–   4.   United States—Social conditions—1945–
I.   Title.
HB3505.E247   304.6   79–56369
ISBN: 0–465–00688–4

265734

To

"P."

CONTENTS

PREFACE   *ix*

I
The Argument

1. The Accident of Birth: Generation Size
   and Personal Welfare   *3*
2. The Economic Fortunes of Young Adults   *15*

II
Family

3. Marriage and Childbearing   *37*
4. Women's Work   *60*
5. Breakdown of the Family?   *79*

III
Society and Economy

6. Social Disorganization   *97*
7. Stagflation   *112*

IV
Implications

8. The Future   *131*
9. Conclusion   *145*

APPENDIX TABLES   *165*
NOTES   *181*
INDEX   *197*

PREFACE

THIS IS THE STORY of my generation, of my children's generation, and perhaps of my children's children's generation. Not the whole story, to be sure, but an important part of it.

I started on this inquiry in the latter part of the 1950s, while studying the causes of the dramatic and unexpected postwar baby boom. As one of the baby boom parents myself, I came increasingly to feel, as a result of my work, that the history of my generation had been importantly affected by its scarcity, the result of the low birth rate era of the late 1920s and 1930s. I came also to suspect that the United States might be involved in a self-generating mechanism, by which low fertility in one twenty-year period led to high fertility in the next, and vice versa—as I wrote in "The American Baby Boom in Historical Perspective" in 1962. In 1968 this earlier work culminated in a historical monograph, *Population, Labor Force, and Long Swings in Economic Growth*.[1]

In this book, I have returned to the story, explicitly adopting a generational frame and broadening the scope to a much wider range of social and economic conditions. I have sought to explain in part the history of recent generations of young American adults, the "good times" social milieu of the decade or two before 1960, and the economic and social malaise that developed thereafter. I have also attempted to look at the prospects for the next twenty years and beyond.

I have tried to make this more readable than the 1968 book, deliberately choosing a less technical style because I wished to reach a wider audience, professional and nonprofessional. The topics range across several social sciences, and I felt that a less technical treatment would help communication with different scholarly disciplines. Also, as I have said, this book is about the experience of recent generations of Americans, and I wanted to make the story accessible to them because it is *their* story.

This book is far from a solo work. First, and most of all, I must thank Eileen M. Crimmins, whose participation in the study from its conception to its birth approaches that of a co-author. I have benefited repeatedly from her ideas, suggestions, and criticisms in lengthy discussions ranging over economics, sociology, and demography. If the resulting analysis is more soundly conceived, as I think it is, it is in no small measure owing to her extensive contribution.

I must also thank Michael L. Wachter, whose own work and interest in this topic encouraged and stimulated me to return to the subject; and Jerry Coombs, who also spurred my renewed interest. Much of the research reported on here was financed by a grant from the National Institute of Child Health and Human Development Center for Population Research (HD–05427) held by Wachter, Robert A. Pollak, and myself, and I have benefited considerably from discussions with both of them. I have also been helped by my colleagues on the Social Science Research Council Committee on Social Indicators, especially James A. Davis, Robert M. Hauser, Kenneth C. Land, Stanley Lebergott, William M. Mason, Robert Parke, and Albert J. Reiss, Jr. Among others who have contributed valuable comments on part or all of this or earlier versions of the argument are Moses Abramovitz, Gretchen A. Condran, Ralph B. Ginsberg, David M. Goldberg, Robert Higgs, Simon Kuznets, Ronald D. Lee, Peter H. Lindert, Karen Oppenheim Mason, Martin O'Connell, Valerie K. Oppenheimer, and Graham S. L. Tucker. The critical comments of my editor, Martin Kessler, have done much to improve the manuscript. Its readability has been considerably enhanced by the work of Diane Bonner and Phoebe Hoss.

Preface

For help in assembling and analyzing statistical data on the wide range of topics covered here and for many constructive comments, I am especially grateful to Morton Owen Schapiro. I have also had the benefit of excellent typing and research assistance from Aline S. Rowens, whose cheerful demeanor under the stress of repeated deadlines was a special boon. A large part of an earlier draft was produced with equal grace and facility by Jenny L. Spancake. Others who have helped in one way or another are Adele B. Burns, Robert D. Cohen, John Daniel Easterlin, Susan P. Easterlin, Lisa M. Ehrlich, William W. Greer, Mahmoud S. Issa, Frank R. Lichtenberg, and Deborah C. K. Wenger.

I

The Argument

1

The Accident of Birth: Generation Size and Personal Welfare

THE BELIEF that one can shape one's own fate—that hard work will be rewarded and laziness punished—has a strong hold in our society. Individual effort is, of course, one factor affecting a person's destiny. But forces beyond the control of the individual also play a role in determining one's life, and for the bulk of the population, they may often play a crucial role. Natural catastrophes are one example; dramatic political events like wars, another; great depressions, a third.

This book is about how and why one such force operates in America today—the comparative size of one's generation. By "generation," I mean a group of persons born in a particular year. The circumstances of one's birth—whether one is born poor or rich—have always played an important part in one's life chances. But over the last four decades, birth and personal welfare have become connected in a new way. It is now becoming clear that in the post-World War II economy the success of a generation's members may be crucially affected by how numerous they are. For those fortunate enough to be members of a small generation, life is—as a general matter—disproportionately good; the opposite is true for those who are members of a large generation.

Preview

In this book, I shall show that the fortunes of a "baby boom" generation, one hailing from a period when the national birth rate is high, will differ in a number of ways from those of a "baby bust" generation, one born in a period when the national birth rate is low:

· The economic fortunes of young workers—their earnings, unemployment experience, and rate of advance up the career ladder—will be adversely affected.
· Young adults will be hesitant to marry, and premarital conceptions will be more likely to end up as illegitimate births.
· Young families will be under disproportionate pressure to put off having children.
· Those young women who choose motherhood will often combine a job outside the home with the care of children.
· Marital strains will be high, and divorce unusually frequent.
· Psychological stress among young adults will be comparatively severe, and suicide, crime, and feelings of alienation unusually high.
· The economic environment during the critical family-forming years is likely to be unfavorable—characterized by "stagflation," a combination of relatively high unemployment and accelerating inflation.

In these and other ways, a baby boom generation finds the going comparatively tough. Correspondingly, a baby bust generation will find life relatively easy. The pressure or the absence of numbers is likely to be felt more in the early years of adult-

The Accident of Birth

hood; but the effect, whether good or bad, is likely to follow a generation throughout its existence. In this sense, year of birth marks a generation for life.

One may argue—doubtless with some truth—that generation size has always been a factor in one's fortunes. In this century, however, new conditions in American society have caused that size to assume unusual significance. Policies of the federal government that have severely restricted immigration and maintained a high and growing level of employment have sharply altered the historical conditions of American labor supply and demand and, since World War II, have resulted in a new relation between population and the economy.

Several of the conditions I will discuss—stagflation, rising crime and suicide rates, accelerating divorce and illegitimacy—have been taken by some observers of the contemporary scene as symptomatic of a general deterioration in the American economy and society. My view leads to reconsideration of this notion. If I am correct, the severity of these conditions will abate over the next two decades as the pressure of numbers diminishes. Then, starting around the turn of the century, there may be a return to conditions like those of the past two decades. As I shall show, we may now be embarked on a long cycle, in which two decades of relatively good times are followed by two decades of relatively poor times, and so on.

There will be those who disagree with what I have to say. This disagreement is partly a matter of misunderstanding: I am not claiming that a generation's size is the sole arbiter of its experience or the whole story behind the varying quality of American life in recent decades. Certainly other important forces have been at work. I am arguing simply that the influence of numbers has been much greater and more pervasive than has heretofore been recognized.

In part, disagreement stems from conflicting views of the causes of some of the specific developments I discuss. These other views may identify additional factors at work, but some of them strike me as questionable. I shall try to take up both kinds of differing views as I go along.

Let me try at once, however, to forestall one possible misinterpretation of my argument. A hasty and technically sophisticated reader might conclude that it contains nothing new—that it is widely recognized that changes in the age distribution of the population have important consequences for national rates of crime, unemployment, births, and so on. For example, the rise in the nation's crime rate since the early 1960s is frequently linked with the increase of young adults in the total population. The reasoning is simple: young adults commit crimes more frequently than do persons younger or older than they. Hence, in a population where young adults are relatively numerous, there will tend to be a higher incidence of crime than in one where young adults are relatively scarce. Because young adults have become relatively more numerous in the United States since the 1960s, the nation's average crime rate has risen.

Such effects of the changing age structure of the population —technically dubbed "age-composition" effects—are important, but they are not my concern. When I state, for example, that an increase in young adults in the population raises crime rates, I mean that it raises the rate of crime among those whose numbers are growing—that is, among young adults themselves. My primary interest, in other words, is in the conditions specific to the group whose number is changing—what are called "age-specific" effects. One will not find here, therefore, the usual discussions of the effect of changing age distribution of the population on such things as school construction, voting, and retirement funding. My concern is with a set of age-structure effects that has gone virtually unnoticed.

Virtually . . . but not wholly. Other scholars have arrived at similar conclusions for a number of specific developments. I am not the only one to point to the age-specific effects of generation size, although I am perhaps the first to stress their pervasiveness. In what follows, I shall often draw on the work of others to build my argument.

Laying the Groundwork

In the remainder of this chapter, I will cover two background matters. One is to help the reader locate his generation on the scale of large and small. The other deals with a fundamental consideration of the analysis—the specialized roles of husband and wife within the family.

How Large Is Your Generation?   Throughout this book I use the term "generation" in much the same way as one would "college class." Those in the "class of '55" are those who graduated in 1955; the "generation of 1955" includes all those born in that year.

A generation, as I define it, is what "demographers"—specialists in the study of population—mean by a "birth cohort"; and I use "generation" and "cohort" interchangeably. Neither term needs to be confined to a single year. Just as we speak of the graduates of the sixties, so too can we speak of the generation or the cohort of a particular period, such as the baby boom cohort of the fifties. My interest is primarily in those born in periods of low or high birth rates, rather than in any one high or low year.

Low birth rate periods produce small generations; high birth rate periods, large generations. The reader can locate himself on the scale of generation size in figure 1.1, a historical record of birth rates. Note especially the low birth rates in the decade of the 1930s, the high levels reached in the period following World War II to 1960, and the recent return to low levels. Those born in the 1930s—the birth cohort of the thirties—

belong to a small generation; those born in the 1950s, to a large one. Note, too, that the movement from low to high birth rates and back again is not abrupt but occurs fairly smoothly. Small generations shade gradually into large, and vice versa. Much of my attention will focus on those at the extremes, those born in the thirties and fifties. The experience of those who hail from periods when birth rates were at an intermediate level would be transitional in nature.

FIGURE 1.1

The Birth Rate since World War I

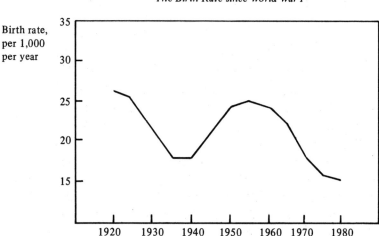

Birth rate, per 1,000 per year

After troughing in the 1930s, the birth rate climbed to a post-World War II peak in the 1945-60 period, and has since declined noticeably again. (The birth rate is the ratio of births in a given year to the average population; annual birth rates are here averaged over five-year periods to bring out the longer term pattern.)
Source: Appendix table 1.1.

At various places, I speak loosely of generation size in terms of large or small "numbers." A careful reader, however, will note that generation size is identified here, not with the absolute number of births in a given year or period, but with the average birth *rate,* the number of births per thousand total population.

The Accident of Birth

Family Roles. I have mentioned that generation size has important effects on family life—on when one gets married, how many children one has, the probability of divorce, and so on. But the effects of generation size sometimes differ between men and women. To understand why, it is important to recognize the different family roles of husband and wife. In most families, husband and wife will in the lifetime of each have responsibilities as a parent within the home and as a worker outside the home. But the relative importance of these roles typically differs: for the man, work outside the home is more important; for the woman, her job as a parent is predominant. And society's judgment on the respective accomplishments of husband and wife is similarly differentiated: a man is judged first on whether he is a "good provider"; a woman, on whether she is a "good mother." This difference in the family roles that men and women expect to play is widely recognized by those who study American society, as illustrated by the following quotation from the work of a leading sociologist, Alice S. Rossi:

[M]en have no freedom of choice where work is concerned: they must work to secure their status as adult men. The equivalent for women has been maternity. There is considerable pressure upon the growing girl and young woman to consider maternity necessary for woman's fulfillment as an individual and to secure her status as an adult.[1]

This view of their primary roles within the family, which most men and women share, is chiefly due to the way that boys and girls are brought up, or, as sociologists would say, are "socialized." [2] Boys are prepared chiefly for the workaday world; girls, primarily for the care of family and home. Parents play a key part in this sex-role indoctrination by providing "role models" for boys and girls. Watching their parents, children learn that the father's work outside the home is all important, while the mother who has a job must sacrifice it for

9

the welfare of the children. Thus, if a child is sick, it is the mother who attends to the child's needs, while the father is excused from this task because of the greater "urgency" of his job. Children also learn sex roles from their parents by the strictures that the latter impose on their behavior ("big boys don't cry," "good girls don't get all messy"), from the toys they are given (science kits versus play stoves), from the way their rooms are decorated, and in numerous other ways. The parents' influence is reinforced by sex typing in schools, church, and the media. Schoolteachers, for example, tend to orient boys toward math, science, and competitive sports to prepare their analytical skills for dealing with a competitive world. Television shows present men as doctors, soldiers, and astronauts; women as nurses, teachers, and, above all, homemakers. Peer-group influences further buttress traditional sex-role images: the girl who is "into" math and science is an oddball and a loner; the cheerleader, popular with boys, is the one to be admired.

In recent years, traditional sex roles have been questioned as never before. And there are real signs of change. Certainly schools are doing more to treat students the same regardless of sex; and businesses, colleges, and other institutions are trying to expand opportunities for women. Also, surveys show that increased proportions among both sexes are in favor of equal labor market rights for women and of jointly making important household decisions.[3] However, on the issue of whether or not there has been a fundamental shift in views among the population about the principal roles that husband and wife should play in the family—the answer suggested by the evidence is negative. Today, as they reach adulthood, most men and women envisage the traditional arrangement in which the man in the family is a full-time worker throughout his life, while the woman drops out of the labor force to have and to raise at home two or more children, at least until they reach school age. The woman is expected to work outside the home before childbearing and also, in most cases, to return to the labor force

The Accident of Birth

after the children reach school age. Moreover, the job the woman expects to hold is usually a traditionally female one, just as is the one the man expects to hold traditionally male. Here are the results of some recent surveys of young adults, the group for whom any significant change that has occurred is most likely to be noticeable (survey dates and age groups vary somewhat because of differences among the surveys):

1. There has been little backing away from the ideal of motherhood for young women. In 1977, three out of every four *single* women aged eighteen to twenty-one expected to have at least two children; among married women in this age group, the proportion was four in five.[4] As demographer Judith Blake has pointed out from studying similar survey responses on ideal family size, despite the large decline in the birth rate, Americans today—including young Americans—"are highly tolerant of large families and noticeably intolerant of the one child family or childlessness."[5]

2. A national sample of high school seniors from the class of 1977 were asked how they felt about different work situations of a husband and wife with preschool children.[6] Out of four possible ratings (not acceptable, somewhat acceptable, acceptable, desirable), seven out of ten considered as not acceptable a situation in which both partners work full time; over half of the remainder gave this situation the second lowest rating (somewhat acceptable). In contrast, the traditional arrangement—husband works full time, wife does not work—received the two highest ratings from four students in five (desirable, acceptable), with the division between the two ratings about equal. Male respondents tended to be more traditional than female in their evaluations, but the difference by sex was slight.

3. In 1979, a national cross section of teenagers aged thirteen to eighteen were asked: "As of right now what kind of work do you think you will do for a career?"[7] The job aspirations of teenagers are likely to be unrealistic, of course, with an

emphasis on glamorous occupations; hence, one cannot take their responses as indicative of the lines of work they will actually pursue. What is interesting, however, is the difference in response between boys and girls. Here are the top ten career choices of each:

Rank	Boys	Girls
1	Skilled worker (mechanic, etc.)	Secretary
2	Doctor, dentist	Doctor, dentist
3	Lawyer	Musician, artist
4	Musician, artist	Nursing
5	Professional athlete	Teaching
6	Electronics	Stewardess
7	Military	Accountant, auditor
8	Business	Lawyer
9	Aviation industry	Social worker
10	Architect	Psychologist

There is some evidence of new aspirations among young women in the appearance in the girls' list of doctor, lawyer, and accountant. A comparison of the two lists, however, shows a substantial difference in the occupational orientation of the two sexes: only three occupations appear on both lists, and the girls' list is dominated by what have been traditional female occupations (secretary, nurse, teacher, stewardess, social worker).

4. Somewhat more realistic are the responses of women aged twenty-one to twenty-four who in 1975 were asked about their future job plans: "What kind of work would you like to be doing when you are thirty-five years old?" [8] Note that the question specifies an age when all the children of most women would be in school and predisposes the respondent to reply in terms of work outside the home. Nonetheless, only slightly more than half (56 percent) actually specified some job plans; 31 percent answered, "Married, keeping house, raising a fam-

ily"; and 13 percent said, "Don't know." Among those who did plan to be working outside the home, the most frequently named jobs were in traditional female occupations. The ten leading choices were:

1. Teacher
2. Secretary
3. Nurse
4. Social welfare worker
5. Practical nurse
6. Hospital attendant
7. Typist
8. Bookkeeper
9. Artist, art teacher
10. Sewer or stitcher

Together these ten occupations accounted for more than half of those with job plans. Again, the impression conveyed is that most young women continue to think along traditional lines.

I have cited this evidence to show why my reasoning in subsequent chapters is based on the view that traditional sex-role attitudes predominate in the population. In adopting this view, I am not saying that it is what ought to be. My interest is to explain, not to prescribe, human behavior; and to achieve this, my analysis must be grounded on a realistic notion of prevailing attitudes.

Among young couples, a crucial factor—possibly *the* crucial factor—affecting such decisions as whether to marry and when and how many children to have is the economic outlook. Hence, my analysis of the impact of generation size starts in chapter 2 in respect to the earnings and the employment experience of young adults. The conclusions of this chapter lay the foundation for the exposition of the many ways in which generation size affects the family: marriage and childbearing (chapter 3), work outside the home among wives, both young and old (chapter 4), and, finally, conditions sometimes viewed as indicators of family breakdown—the probability of divorce and the incidence of illegitimacy (chapter 5).

These developments in family life, although largely concentrated among young adults, bear on the general well-being of the economy and of society. Chapters 6 and 7 pursue this broader concern, describing the impact of generation size on conditions such as crime, suicide, and stagflation. Chapter 8 shifts to the future, both near term (the next two decades) and longer term (the twenty-first century). Chapter 9 ties the argument together.

2

The Economic Fortunes
of Young Adults

Bliss was it in that dawn to be alive,
But to be young was very heaven! [1]

AND SO IT WAS for the young adults of the fifties, those fortunate ones born in the low birth rate era of the 1930s. Not so lucky are their children.

In this chapter, I shall show how those who first see the light of day in a low birth rate period may look forward to a buoyant job market when they reach working age. Job openings will be plentiful, wage rates relatively good, and advancement rapid. Conversely, high birth rates portend a relative labor market surplus and correspondingly unfavorable effects on one's economic life chances. Two steps are involved in the argument: (1) the effect of birth rates on relative numbers when working age is reached, and (2) the effect of relative numbers on earnings and employment experience.

The economic experience of young adults is considered first because it plays a critical role in shaping so many of the major decisions of young couples. The experience of young men is placed in the foreground because of their primary breadwinner role, but that of young women is much the same, as we shall see. The early working period, however, is neither the first nor the last time that generation size makes its mark. Hence, I note various ways in which generation size exerts its effects throughout the life cycle, from childhood to retirement. Finally, I consider why generation size has since 1940 played a much more important role in shaping one's economic life chances than it did before.

Birth Rates and Relative Numbers at Working Age

If you are a member of the low birth rate generation of the 1930s, then when you reached adulthood you entered a labor market in which younger workers were in short supply compared with older. If you belong to the high birth rate generation of the 1950s, then you encountered a labor market in which younger workers were relatively plentiful.

To see how low birth rates produce a relative scarcity of younger workers when working age is reached, and vice versa, consider first the connection between *number* of births and *number* reaching working age. Between birth and entry into the labor force, the number in a generation will be depleted by mortality and may be supplemented or depleted by migration. But if the effects of mortality and migration do not differ very much from one generation to the next, as has been true in recent decades, then differences among generations in their number when they reach working age will depend very largely on differences in their number at birth. Hence, ups and downs in the number of births imply corresponding ups and downs about twenty years later in the number of labor market entrants.

The Economic Fortunes of Young Adults

The delayed effect of number of births on number of labor market entrants was noted in a report by Denis F. Johnston, director of the Social Indicators Project in the Office of Management and Budget. Johnston points out that from 1953 to 1963 the labor force was increasing by about 880,000 per year, but from 1964 to 1974, when the baby boom generation began entering the job market, the annual increase of the labor force nearly doubled—to about 1,740,000 per year.[2]

Just as variations in the absolute number of births have a delayed effect on the absolute number of labor market entrants, so too do variations in birth *rates* have a delayed effect on the *relative* number of younger workers—the proportion of younger to older. The birth rate itself can be viewed as a proportion of younger to older persons. Thus, the birth rate in 1935 is the proportion between those born in that year to the total population at that date, all of whom are as old or older. Consider now the ratio in 1955 of those twenty years old to all those twenty years and older. This ratio will not be exactly the same as the birth rate of 1935 because of the effects of mortality and migration on both the numerator and the denominator. But as I have noted, these effects have not differed that much from one generation to the next. Hence, *differences* among generations in this ratio will be largely a reflection of differences in initial birth rates. A protracted *period* of low birth rates, as in the 1930s, and of high birth rates, as in the 1950s, has therefore resulted, with a two-decade lag, in corresponding periods of shortage and surplus of younger men compared with older.

To illustrate this, figure 2.1 compares birth rates since 1920 with the proportion of younger to older men twenty years later (the picture for women would look virtually the same). In keeping with common practice, the male working-age population is taken as roughly between ages fifteen and sixty-four. Using age thirty as a dividing line, one can divide this popula-

FIGURE 2.1

The Lagged Effect of the Birth Rate on
the Relative Number of Younger Men
Compared with Older

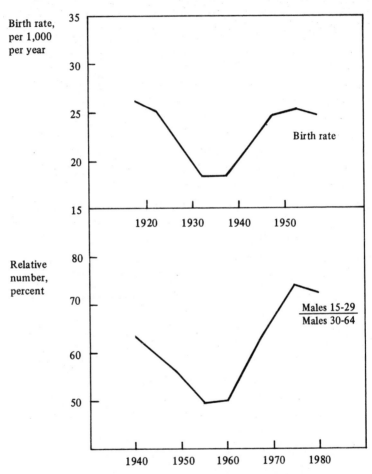

Swings in the birth rate, shown at the top, are echoed about
two decades later in the proportion of younger to older men
in the working-age population, the bottom curve.

Source: Appendix Tables 1.1 and 2.1.

The Economic Fortunes of Young Adults

tion into "younger" and "older" segments. Although any fixed division is arbitrary, age thirty seems reasonable because the futures of most young men are largely set by the time they reach that age.

The bottom curve of figure 2.1 shows the proportion of younger to older men at various dates since 1940. Over the period as a whole, there were in the working-age population, an average of five men under thirty for every eight men thirty and over. Note, however, the marked variation in this proportion during the course of the period. From a value in 1940 slightly above average, the proportion dips to a trough in 1955–60, when there were two younger men for every four older, and then rises to a 1975 peak of three younger for every four older. That this swing in the proportion of younger to older men is due chiefly to a swing in the birth rate twenty years earlier can be seen by comparing the curve in the lower panel with that in the upper. In effect, the relative supply of younger workers is echoing the birth rate two decades earlier.

The Effect of Relative Numbers on a Generation's Economic Success

Thus, the low birth rates of the 1930s meant that the babies born in that period had as young adults enjoyed a labor market in which they were unusually scarce compared with older workers. The high birth rates of the 1950s have in recent years produced a situation in which younger workers are exceptionally numerous compared with older. In turn, a generation's relative scarcity or abundance when its members are launching their careers has an important effect on their economic progress. If you came from the low birth rate generation of the

1930s, on reaching working age you experienced relatively little unemployment, were able to find a job at good wages fairly easily, and advanced up the career ladder rather quickly. If you are from the baby boom generation of the 1950s, the opposite is unfortunately true. Here are some excerpts from *The New York Times* that illuminate the sharp contrast between the 1950s and the 1970s in the labor market of younger workers:

1950s	*1970s*
Job Market Found Good for Youths	**Youths March in Capital, Seeking Creation of Jobs**
May 20, 1955—This year's high school graduates got word today from James P. Mitchell, Secretary of Labor, that job prospects are good.	April 9, 1978—Several thousand young people, many of them out of work, demonstrated on Capitol Hill today, saying that the President and Congress have not done enough to create jobs.
More Jobs Offered to the College Man	**The Art and Science of Finding a Job**
April 30, 1955—Job opportunities and starting salaries for the June college graduate are hitting an all-time high. A nation-wide survey showed today that representatives of business and industry already were on the campuses to interview applicants.	June 18, 1978—Finding work on [Long] Island is proving to be a pressing problem, not only for the printer or laundress, but also for many others with backgrounds in a myriad of fields—especially the college graduate.
	Alas, the college graduate attempts to enter the Island's tight labor force with credentials but lack of work experience. Then the graduate gains another kind of experience: that of outright rejection or being put on "active files."

The Economic Fortunes of Young Adults

Industry Woos College Student and Borrows Faculty Talents— Recruiter Seeks Reservation on Campus a Year Ahead—Salaries Still Rising

Mental Health Centers Booming as College Competition Rises

November 25, 1956—Industry's manhunt is under way. Placement offices at college campuses across the nation are the scene of feverish activity. Talent scouts for leading companies are already making their sales pitch for the services of June graduates. . . . Most companies are putting more recruiters in the field and increasing the number of campuses they visit. . . . Companies are pursuing long-range programs. When personnel men visit a campus this year, they make reservations for 1958. Failure to do this would cause them to be frozen out completely, or to be given an undesirable date.

May 27, 1978—The anguish of increased competition for grades, jobs, and graduate school admission is weighing heavily on the nation's college students, and a sign of their desperation is the booming business at the mental health counseling centers that have become fixtures on so many campuses. Such centers, staffed by psychiatrists, psychologists, and social workers, are now integral in helping students cope with academic life and recognize the realities that await them in a world in which they may have to reset educational and career goals.

Why should this be so? Why should relative scarcity have a favorable effect on a generation's labor market experience, and vice versa?

The answer is, simply put, supply and demand. To elucidate, imagine that the proportion between the jobs available for younger workers and those for older workers is constant: as the economy expands, these jobs grow at about the same rate— that is, the demand for younger and older workers grows equally. Imagine, on the other hand, that the supply of younger workers compared with older workers changes noticeably over a period of a decade or so. At a time when the relative supply of younger workers is high, competition among them will

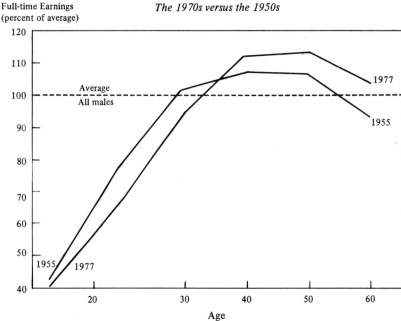

FIGURE 2.2

*The Earnings of Younger Men
Compared with Older:
The 1970s versus the 1950s*

Full-time Earnings
(percent of average)

For each date, earnings at the age shown at the bottom of the chart are expressed as a percentage of the average for all ages (the horizontal broken line). At both dates, the left-hand portion of the curve lies below the average, showing that younger workers earn less than older. In 1977, however, the shortfall for younger workers is greater than it was in 1955, showing that the relative position of younger workers is worse at a time when their relative numbers are greater.

Source: Appendix Table 2.2.

The Economic Fortunes of Young Adults

be intense, and employers can be choosy. Younger workers may have to take considerable time and effort to find satisfactory jobs, salaries may be disappointing, and advance up the career ladder may be frustratingly slow. Conversely, when younger workers are in short supply, employers find themselves competing, while younger workers pick and choose. To attract needed workers, employers will be much more likely to snap up those seeking jobs and to offer higher wages.

Stories such as those just quoted from *The New York Times* are perhaps the best testimony to the personal experiences of those who were young adults in these different periods. But we also can read the story of demand and supply in statistical averages that summarize the earnings and employment experiences of these individuals. Because older men are further up the career ladder than younger, their earnings are typically above average and younger men's are below. However, when the number of younger workers grows relative to older, the wages of the younger fall even further below the average while those of older rise further above. This is shown by the curves in figure 2.2.[3]

I want to emphasize that I am talking about the *relative* earnings of the young. In absolute terms, full-time workers in their early twenties in 1977 earned almost one-third more than their counterparts did in 1955, even after adjusting for the sharp increase in the cost of living. The situation of the young has deteriorated relative to older workers. But as the following chapters will show, relative income is critical in determining the behavior of young adults as well as their feelings of well-being.

The deterioration in the relative earnings of young men with the growth in their relative numbers is repeated in unemployment rates. A higher unemployment rate for younger workers than for older is normal and reflects their newness in the job market, their job-seeking activity, the tentativeness of their job commitments, and so on. In the 1970s, however, the

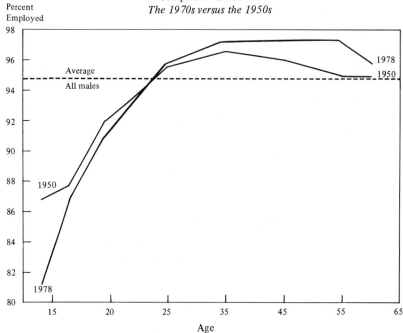

FIGURE 2.3

*Employment Rates of Younger Men
Compared with Older:
The 1970s versus the 1950s*

Percent
Employed

At both dates, the left-hand portion of the curve lies below the average (the horizontal broken line), showing that younger workers have lower employment rates than older (that is, their *un*employment rates are higher). In 1978, however, the shortfall for younger workers is greater, showing that the relative position of younger men is worse when their relative numbers are greater. (The dates chosen here for comparison differ slightly from those in figure 2.2 because for this chart the average unemployment rate for all males is the same.)

Source: Appendix Table 2.3.

The Economic Fortunes of Young Adults

unemployment rates of younger workers compared with older were noticeably worse than in the 1950s (see figure 2.3). Thus, the weight of numbers sharply aggravates the relative unemployment as well as the relative earnings disadvantage of the young. It is as though young and old were at opposite ends of a seesaw, with the ends corresponding to their earnings or employment levels. The seesaw is always tilted against the young. But when relatively more young workers are piled on their end, the seesaw tilts even more against them.

As I have mentioned, these changes in younger men's relative earnings and unemployment rates reflect shifts in the supply of young versus old relative to normal demand. If wages were highly flexible, the effect of changes in relative numbers would be confined to rates of pay. But minimum wage laws, unemployment compensation, and similar conditions limit downward pressure on wages, so the effect of their numbers shows up both in earnings and in unemployment rates.

Job Advancement. Most men are on a "career ladder." Blue-collar workers may start in unskilled or semiskilled jobs requiring little experience and then advance to positions as craftsmen or foremen. Junior executives move up to top management positions. The more rapid the rate of advance, the more one's economic fortunes prosper.

Starting positions on the career ladder are sometimes called "entry-level" jobs. Indeed, the career ladder stretches back into the educational system since a variety of entry-level jobs, both blue and white collar, require specialized vocational or professional training.

As a general matter, entry-level jobs tend to grow at about the same rate as the economy's overall demand for labor. A big expansion in the supply of younger men compared with older has its first impact on entry-level jobs and on the educational institutions that feed graduates into such jobs. The difficulty of admission to medical, business, and law schools or, more generally, to "the college of one's choice" has been a widely publicized problem of today's surplus cohorts. But the competition at this initial stage is only a harbinger of things to come. Those fortunate enough to get a hand on one rung must hold tightly and reach out carefully for the next, for many others are trying to shoulder them aside. This means that, on the average, a surplus generation will move up the career ladder more slowly than a scarce generation. A generation's numbers thus affect not only employment and pay rates but the pace of promotion as well.

Demand versus Supply Changes. I have stressed how changes in the relative supplies of younger workers affect their economic fortunes. Theoretically, increased demands for younger workers might offset differential changes in labor supplies. If, for example, occupations in which young persons have a disproportionately large share, such as operatives, laborers, and clerical workers, grew at much higher rates relative to the average after 1960 than before, this would generate a favorable change in the demand for younger workers, which would tend to compensate for the adverse change in their supply. As it turns out, in this respect, there was virtually no difference between the two periods, and hence, the notion of changes in demand offsetting supply can be safely dismissed.[4]

Can Younger Men Do Older Men's Jobs? If younger male workers could substitute easily for older, then the story would be different. A relative abundance of younger men could simply shift into older men's jobs without any adverse effect on their relative earnings and unemployment. But younger men are not good substitutes for older. Older workers have usually acquired more skills than younger. They also have greater experience with the firm in which they are employed, which adds to their productivity, and they are therefore viewed by employers as more reliable and responsible.

Hiring practices also reduce the possibility of younger workers substituting for older. Personnel departments, for example, may include age or age-related characteristics, such as experience, as requirements for certain jobs. Hence, young applicants may be viewed as unqualified for older men's jobs. Unions may urge employers to give preference to members with seniority in selection for jobs. Such practices reinforce the fact that younger men cannot substitute for older men in the labor market.

A reduction in the relative wages of the young caused by their increased relative numbers may encourage employers to substitute younger for older workers. But the extent to which wages can fall is limited by minimum wage laws. Also as wages drop toward the level of unemployment compensation, some workers will discontinue looking for work, and their withdrawal from the job market will further reduce the competitive pressures that reduce wages. Moreover, employers are less likely to find it worthwhile to make the adaptations necessary to substitute young for old when the shift in relative numbers is transitory (as has been true since 1940) than when it is longer term. Hence, both the wage inducement to substitute younger for older workers and the employers' response to it are likely to be limited. Therefore the argument that younger workers can readily substitute for older can be set aside.[5]

Women's Labor Market Experience. So far I have focused on younger men. The economic experience of younger women parallels that of the men, but generation size has an even greater adverse impact. For example, in the "small generation" year of 1955, the earnings of women twenty to twenty-four who worked full time all year were almost the same as those of women aged forty-five to fifty-four; by the "large generation" year of 1977, the earnings of the younger group had fallen to 85 percent of those of the older (see appendix table 2.2). Unemployment rates are similar: between 1950 and 1978, the unemployment rate of females twenty to twenty-four rose from 6.9 to 10.1 percent; for females forty-five to fifty-four, the rate *dropped* from 4.5 to 4.0 percent (see appendix table 2.3; the dates here used for comparison are the same as those used for men in figures 2.2 and 2.3).

As I have mentioned, the trend from the 1950s to the 1970s in the absolute amount of earnings was generally upward, even after adjusting for the rising cost of living. The deterioration in the earnings of younger workers relative to older implies that the rate of increase in earnings was lower for the younger group. Within that group, however, the rate of growth was even less for women than for men; hence, the earnings of young women deteriorated not only relative to older women but also relative to younger men. In 1955, among full-time year-round workers aged twenty to twenty-four, females' earnings were 84 percent of males'; by 1977, the figure had dropped to 77 percent. This decline in the relative position of women was not confined to younger persons—it applied also to older workers. Moreover, it is repeated in the movement of unemployment rates. From 1950 to 1978, the rise in unemployment rates for younger women is greater than that for younger men; for older women, the decline in unemployment rates is less than that for older men.

In explaining the relative deterioration of younger workers' earnings and unemployment experience over the last two decades, I have stressed the growth in their relative numbers in the labor market. Differential growth in relative numbers also seems to bear on male-female differences—a worsening of women's earnings and unemployment experience relative to men's has been accompanied by a growth in their relative numbers. Indeed, if one divides the labor force into four age-sex groups —younger males, older males, younger females, and older females—the ranking of the groups according to the relative change in earnings and unemployment experience is the opposite of that according to the change in relative numbers— older men, whose relative numbers have changed least, have done the best; younger women, whose relative numbers have grown most, have done the worst (see, for example, appendix table 7.1).

I have said that generation size is the principal explanation of the growth in younger versus older workers. In the growth of women workers relative to men, of course, a number of other factors are at work. As we shall see in chapter 4, however, generation size has a part in this picture as well—the unusually large growth of younger women working outside the home is partly a consequence of increased generation size. But more on this at that point.

The decline in the relative earnings of younger women partly reflects an adverse shift in their occupational distribution between the fifties and seventies, that is, an increase in the proportion working at lower paying jobs. This is suggested by studies that show that for the same type of work there has been little change—for better or for worse—in the earnings of women compared to that of men.[6] But this adverse shift in the types of jobs held by younger women is itself a result of their relative numbers. In the 1950s, because of their scarcity in the labor market, younger women could pick and choose among jobs and could avoid lower paid types of work.[7] This was not true in the seventies, when the relative supply of younger women was vastly greater.

Generation Size over the Life Cycle

I started my story of generation size and personal welfare by focusing on the economic experience of young adults because this is so vital to many of the major decisions that young couples make. But the effect of generation size, good or bad, persists throughout the life cycle. Every generation follows the pattern of below-average earnings in early working life and above-average earnings later. But the earnings pattern of a small generation is more favorable throughout its career than that of a large generation—during early working life, the earnings of a small generation are not so far below average; at mid-career, earnings are further above average. The same applies to unemployment experience. Although both large and small generations have higher unemployment rates in early working life than in later, the unemployment rate of a small generation is less at each stage of its working career than that for a large generation.

This can be seen in figures 2.2 and 2.3, if one recognizes that in each figure the twenty-year-olds on the 1950s curve are the same generation as the forty-year-olds on the 1970s curve. In both figures, the comparative position of this low birth rate generation is clearly superior at both stages of the life cycle. Thus, the effect of generation size follows a cohort through its working career.[8]

But entry into working life is not the first time that generation size makes its mark. In the course of their upbringing, both large and small generations have had significant clues as to what the future holds. For large generations, the typical story is one of crowding; for small generations, it is one of nurturance and special attention. At home, for large generations the pressures of sibling rivalry are high, and mom and dad are hard pressed to give time to each child's physical and emotional needs. At school, classrooms are overflowing, and inexperienced teachers, aided sometimes by volunteer assistants, are likely to be one's fate. Competition is everywhere severe—in class, in extracurricular activities, for part-time jobs. Because of the pressure of numbers, entry standards are likely to be raised. Experience will be required for jobs where previously on-the-job training was the norm (but where is one to get experience?); admission standards to the "best schools" are likely to be raised. For example, when the vanguard of the baby boom generation, those born in 1946, hit the colleges, they

The Economic Fortunes of Young Adults

were confronted with elevated SAT requirements for admission; today, as the "birth dearth" of the 1960s starts to take its toll on the number of college applicants, admission standards are being relaxed.

The point of all this—that its size marks a generation for life—has been eloquently made by H. Scott Gordon, another economist who has put his finger on the connection between generation size and personal welfare. Speaking of those born in Rumania in the late sixties, when birth rates shot up to unusually high levels, Gordon says:

These poor souls came into a crowded world—crowded by themselves. There was crowding in the maternity wards when they first saw the light of day; there was crowding in the kindergarten classes when they entered the school system in 1971; there will be crowding in the universities in the mid-1980s; crowding in the search for jobs and housing a few years later; and so on until there is crowding in the funeral parlors and the cemeteries. Rumania may not be over-populated in the aggregate but the infants of 1967 will live all their lives in a crowd. It was not an optimum time to be born, in Rumania.[9]

In contrast, Gordon invites one to "consider the fate of one who is born in [a] sheltered [low birth rate] trough," preceded and followed by longer periods of high birth rates:

When he opens his eyes for the first time it is in a spacious hospital, well-appointed to serve the wave that has preceded him. The staff is generous with their time, since they have little to do while they ride out the brief period of calm until the next wave hits. When he comes to school age, the magnificent buildings are already there to receive him; the ample staff of teachers welcomes him with open arms. In high school, the basketball team is not as good as it was but there is no problem of getting time on the gymnasium floor. The university is a delightful place; lots of room in the classes and residences, no crowding in the cafeteria, and the professors are solicitous. Then he hits the job market. The supply of new entrants is low, and the demand is high, because there is a new large wave coming behind him providing a strong demand for the goods and services of his potential employers. His parents tell him how tough things were when they entered the job market, but he has many offers, and good ones, so perhaps he takes a year or two off to go backpacking in Europe—a good idea, provided that he doesn't stay too long. Having returned from his *wanderung,* he takes a job with the Ajax Tool and Die Company as personnel officer, where he finds that he can meet the firm's need for staff with such ease and effectiveness that work is a pleasure. Finally, he retires, and shortly thereafter the Congress raises social security benefits by a generous sum, finding that it can do so without raising payroll taxes because of the increased numbers in the labour force, and the decline in the number of beneficiaries. He is truly demographically lucky.[10]

The Dawn of a New Era

One might suppose that generation size has always played an important role in shaping personal welfare. But this was not always so, at least so far as economic success is concerned. Before World War II, both the supply of and the demand for labor were much more variable than they are today, and the effect of generation size was dwarfed by other influences. It took new government policies stabilizing demand and capping immigration to set the stage for the leading role that generation size today plays in determining a cohort's economic welfare.

Life before World War II. Consider the fate of a small generation in the more distant past.[11] It might reach the labor market eager and hopeful, but the economy might be in the throes of a deep depression with jobs hard to find. Such was frequently the case before World War II, when the demand for labor fluctuated from one decade to the next. The 1930s, when for a full decade almost one worker out of five was out of work, are but the most recent example—albeit an extreme one—of the long succession of major depressions that dotted the American experience in the century before World War II. If labor demand is grossly deficient, scarce numbers will not benefit a cohort at all.

But what about boom periods? A small generation, presumably, would do especially well if it arrived on the labor market when demand was high. The catch here is unrestricted immigration. Before the mid-1920s, if domestic demand and supply produced a tight labor market—plentiful jobs at good wages—the flow of immigrants from abroad was encouraged. The potential benefit of small generation size was therefore lost in the influx of European workers. This is shown by the historical record, in which large surges in unrestricted immigration match on a one-to-one basis the great booms in the economy.[12] Hence, the bright prospects of a small cohort that came along in a boom period were swamped by competitors from abroad.

The Break with the Past. Federal laws have drastically altered the historical relations between labor demand and supply. On the supply side, federal legislation since the 1920s has sharply restricted immigration in periods when labor demand is high. On the demand side, the federal government, through the Employment Act of 1946, committed itself to maintaining a high and growing level of labor demand through monetary and fiscal policies. The government was aided in accomplishing this by the substantial rise in the relative importance of federal expenditures in the post-World War II economy as compared with those of the prewar economy. These new policies meant that a small generation would not be confronted with a major depression or a rush of foreign competitors. For the first time, generation size could play a central role in one's economic life chances.

Of course, government policy has not entirely eliminated fluctuations in labor demand. But the contrast between recent experience and that of the pre-World War II period is striking. Between 1942 and 1974, the annual unemployment rate rose above 6 percent in only two years, and even then by less than one percentage point. Although knowledge of unemployment rates in the century before World War II is imperfect, it is safe to say that any prior thirty-year period would be characterized by a severe economic depression, with unemployment rates of at least 10 percent or more.[13] Thus, compared with past periods of comparable length, the years since World War II have been marked by a relatively uninterrupted growth in labor demand.

Government policy, moreover, did not wholly eliminate immigration. The record since World War II shows a steady inflow, averaging in the last decade about 400,000 persons per year. In addition, in recent years, a noticeable and much publicized flow of immigrants has been entering illegally. Even adding a reasonable estimate of this to the legal flow, however, would yield current rates of immigration relative to the working-age population that are substantially less than in the past.[14] Moreover, the majority of illegals do not compete with native Americans. A large proportion are filling jobs, such as migrant agricultural workers, that few native Americans care to take.

Although immigration and fluctuations in labor demand have not ceased, their magnitudes, compared with the past, are much smaller. In contrast, as will be seen in chapter 8 (especially figure 8.2), the magnitude of swings in the birth rate and thereby in the proportion of younger to older men is strikingly greater than in the past. The result has been a major shift in the comparative roles of these factors in shaping individual experience. For the first time, the national birth rate at the time of one's birth has come to play a major role in one's economic fortunes.

II

Family

3

Marriage and Childbearing

JUST AS stockmarket analysts anxiously track the course of the Dow-Jones Index, demographers follow the American birth rate. On demographic charts, 1968 marked a historic "breakthrough," as the birth rate plunged below its previous floor, reached in 1933, to 17.5 per thousand population, an all-time low in American history. Since then, the birth rate has dropped even further, reaching in 1976 a new low of 14.7 per thousand.

The recent decline in the crude birth rate is due partly to changed marriage patterns, as young adults marry less and at later ages. Much more important, however, it is due to a decrease of childbearing within marriage. These twin developments, deferred marriage and dramatically reduced marital fertility, have been taken by some as heralding a new age in the evolution of the American family. Charles F. Westoff, a distinguished demographer and former executive director of the staff of the President's Commission on Population Growth and the American Future, cites the steady decline since 1960 in the proportion of young women marrying as perhaps "the unrecognized beginning of a radical change in the family as we know it. Even more important [is] . . . the changing status and role of women in our society . . . the very real changes in women's attitudes toward work . . . marriage, and childbearing. . . . [In addition], the technology of fertility control has improved tremendously." [1] Because of these factors, Westoff argues, "fertility in the United States and other developed countries seems destined to fall to very low levels, probably below replacement." [2] If additional authority is needed, we also have anthropologist Margaret Mead's pronouncement that "American marriage and birth rates will continue to drop in the next quarter century because of changing sex roles and expectations of marriage." [3]

At first glance, such views seem plausible. Yet a closer look makes one uneasy. In 1948, for example, a statement reflecting the consensus among demographers about the future of American fertility sounds uncomfortably like those I have just quoted: "No one anticipates the restoration of levels of fertility that could be regarded as high in a world setting. The range of uncertainty is between rates somewhat below permanent replacement and rates slightly above such replacement." [4] Yet the same year that this statement was made, 1948, also saw the postwar birth rate soar to almost 25 per 1,000. At first, this baby boom was discounted as temporary—the making up of childbearing deferred because of war. But as high birth rates persisted into the late 1950s and as age at marriage continued to fall among persons who had only been children during World War II, it was acknowledged that this explanation did not hold. No persuasive alternative has since been advanced, and this upsurge in postwar marriage and fertility has stood like a warning sign to those who believe that the road to the future is a straight-line continuation of the past.

In this chapter, I shall argue that *both* the postwar baby boom and the subsequent baby bust were in large part the products of swings in generation size that affected the economic circumstances of young adults. Because of their exceptionally favorable economic situation, those from the small generation of the 1930s tended to marry earlier and have more children; the relatively unfavorable economic situation of the large generation of the 1950s made for later marriage and reduced childbearing.

Marriage, Childbearing, and Generation Size

Determinants of Marriage and Fertility. I believe that an important factor affecting a young couple's willingness to marry and to have children is their outlook for supporting their material aspirations. If the couple's potential earning power is high in relation to aspirations, they will have an optimistic outlook and will feel freer to marry and have children. If their outlook is poor relative to aspirations, the couple will feel pessimistic and, consequently, will be hesitant to marry and have children.

Many other factors, of course, affect decisions to marry and have children, such as love for children and religious beliefs. And some couples do nothing to control their fertility; they just go ahead and have babies, whatever the circumstances. My concern is not with everything that enters into a couple's decision to have children or that differs from one couple to the next. My interest is in identifying the particular circumstance (or circumstances) common to many couples that has changed over the post-World War II period and that has resulted first in a surge in fertility and then in a contraction. I do not claim that a couple's economic outlook is the only factor in their decisions to marry and bear children, or even that all couples make conscious decisions about how many children to have. I simply maintain that the economic outlook is one factor common to many couples' decisions and that since 1940 important changes in this factor have caused dramatic swings in fertility behavior.

Note that two elements enter into the judgment about the

couple's economic prospects. One is their potential earning power; the other is their material aspirations. The proportion between the two is what determines judgments on the ease or difficulty of forming a household, and this ratio can vary because of changes in the numerator, denominator, or both. Thus, an optimistic outlook may arise from exceptionally high earnings prospects for the couple, unusually low material aspirations, or a combination of the two. Let us look at each in turn.

Factors in the Earnings Outlook. How does a young couple judge their earnings prospects? Clearly, many considerations are involved, and they vary from one couple to the next, such as an individual's energy, ambition, education, "connections," and so on. Moreover, however long the list one might think up in advance, the actual experience of "working and getting" will dominate in judgments on the earnings outlook; that is, the ultimate test is the labor market itself. For most young adults, an interval of several years lies between starting work and marrying. A Department of Labor study shows that labor market knowledge among young adults is positively associated with years of exposure.[5] Thus, valuable information is accumulated over a period of some length that provides an important basis for projecting the future. If jobs are easy to find, wages good, and advancement rapid, the future will look rosy; if times are bad, the opposite will be true.

The Formation of Material Aspirations. While the labor market may be the principal teacher of earning prospects, one's family of origin is the principal instructor of life-style. By life-style, I mean how the material standards of young adults are formed—why one generation, say, views a car as a luxury and the next, as a necessity. My argument is that the material expectations of young adults are largely the unconscious product of the environment in which they grew up. In other words, economic aspirations are unintentionally learned, or "internalized," in one's parents' home. And this environment is very

Marriage and Childbearing

largely shaped by the economic circumstances, or income, of one's parents. Thus, a child raised in an affluent suburban home, who is driven in an automobile to school, stores, friends, movies, and so forth views the automobile as an integral part of life.

One may cite, of course, a number of other factors affecting aspirations—religious training, formal education, neighborhood environment, peers, and relatives—the multitude of circumstances that sociologists call the "socialization experience," that is, the long years of transition from being a young protected child to a functioning, independent adult member of society. But many of these circumstances are determined by one's parents' income. Nor should style fads be confused with trends in material aspirations. The current craze for casual dress, epitomized by the international market for Levis, is not a rejection of material affluence by the young, as one look at the price tags on Levis quickly testifies. Today's youth may prefer a more casual style, but not a less costly one—witness the cost of such "necessities" as stereos, vans, and rock concerts.

Two young adults contemplating marriage come, of course, from two different families. To predict their material aspirations would require, one might suppose, combining the incomes in their respective families of origin and averaging them, perhaps in proportion to the male's and female's respective weights in shopping decisions. But because people generally find mates from within the same economic class, the income of either family usually approximates the average of the two. For this reason, and to simplify matters, I shall confine myself to the family of origin of the prospective husband.[6]

Relative versus Absolute Income. A couple's assessment of their earning potential might be thought of as their "absolute" income outlook. I have argued that this factor is important in decisions that involve long-term economic commitments. But absolute income is not the controlling factor. The same absolute income may look quite different to two couples with substantially disparate "economic" socialization experiences. To one couple, from an affluent background and with consequently high material standards, $20,000 may leave them feeling pinched; to another, from an impoverished background and with low material desires, it may look like easy street. The same argument would apply to a comparison between two different points in time. To a second generation, $20,000 may not mean so much as it had to the first generation—even if its purchasing power were the same—because the second generation comes from wealthier backgrounds and consequently has formed more ambitious material expectations. Using Gallup poll surveys that asked the question, "What is the smallest amount of money a family of four needs to get along in this community?" Lee Rainwater found that, expressed in dollars of the same purchasing power, the amount increased by one-third between 1954 and 1969.[7] Why did the later respondents think that so much more was needed? The answer is clear: the second survey reflected the respondents' experiences with more affluent conditions, and these experiences elevated the standards by which they judged the minimum amount necessary to get along.

This reasoning can be distilled into a fairly simple notion of "relative" income, which, as the relation between earnings and aspirations, can be defined in simple ratio terms as:

$$\text{Relative income} = \frac{\text{Earnings potential of couple}}{\text{Material aspirations of couple}}$$

To simplify matters, I propose to approximate this by:

$$\text{Relative income} = \frac{\text{Recent income experience of young man}}{\text{Past income of young man's parents}}$$

In this formulation, the recent income experience of a young man in the labor market is taken as shaping the assessment of the couple's earnings prospects; the past income of their

Marriage and Childbearing

parents as establishing the material environment in which they were raised and, thus, their material aspirations. In the numerator, I focus on young men because in most families the man's income accounts for the major share of the total over the life cycle. Also, as we have seen in chapter 2, the economic fortunes of men and women are affected similarly by generation size, with women, if anything, being more adversely affected by large generation size. Hence, the man's situation provides a conservative picture of how a couple's prospects are affected by the size of their generation.

Relative income is not the same as that discussed in chapter 2. There, the income of young men was compared with that of older men at the same point in time. Here, the income experience of young men is compared with that of older persons during a somewhat earlier period because we are now comparing the income of young men with a couple's aspiration levels. These aspirations depend not so much on their parents' present income (although this may be of some relevance) as it does on their parents' income some years earlier, when they were living at home and growing up.

Testing the Relative Income Theory. The argument so far can be summarized quite simply: as the relative income of young adults rises, they will feel less economic pressure and hence freer to marry and have children; as their relative income falls, they will feel increasing economic stress, and marriage and fertility will decline.

This reasoning may seem plausible, perhaps even obvious. But many persuasive arguments can be made about the causes of fertility, particularly about its decline in America since 1960 —as mentioned at the beginning of this chapter. The question is: How does one choose among such arguments?

Fifty or one hundred years ago, this question would have been answered in terms of such things as intuitive appeal, logical consistency, or the rhetorical skills of an argument's proponents. As the standards of natural science have been increasingly applied in the social sciences, however, the answer has become by appeal to the evidence, particularly quantitative evidence. Thus, each reason advanced for the phenomenon is viewed as a hypothesis and is tested against appropriate data. Naturally, the social sciences cannot perform controlled laboratory experiments. But increasingly sophisticated techniques of measurement and statistical inference have been devised to approximate experimental conditions. No attempt is made here to utilize these techniques, known as econometrics, because the requisite data are scarce. But at least a minimal test can be made of the "relative income hypothesis." If the decline in fertility since 1960 is due in substantial part to a decline in the relative income of young adults, then a measure of relative income since 1960 should show a decline similar to that in fertility. In like manner, for the period of the baby boom, a relative income measure should rise with fertility. Is this in fact the case?

A Measure of Relative Income. To make this test, we need a measure of the relative income over the last four decades corresponding to the definition on page 42. However, it is not easy to find the data needed, despite the simplicity of the idea. Ideally, information on the full income history of both young men and their parents would enable us to experiment with a variety of relative income measures. Such experiments, unfortunately, are a luxury. The requisite income history data do not exist, and the income data that are available permit only the crudest approximation to the ideal, and then only for the period since 1957. I will not go into the calculation of the relative income measure presented here—for those who are interested, the source is given in appendix table 3.1—but will merely state that it attempts to capture the basic idea.

In figure 3.1, the bottom line shows the trend in the numerator of the relative income ratio; the middle line, in the denominator; and the top line, the ratio between the two, that is, relative income itself. Income is adjusted for the increasing cost of living over time, that is, in terms of dollars of constant

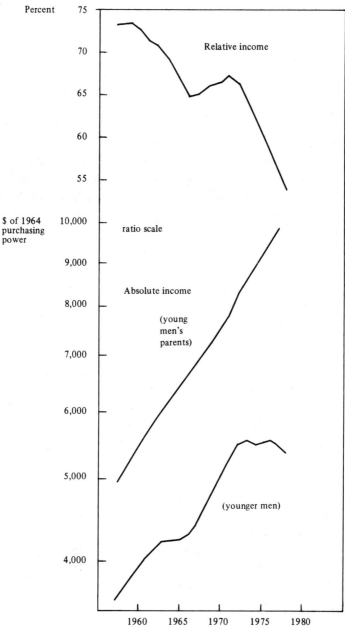

FIGURE 3.1

Measuring Relative Income

Relative income is estimated as the ratio of young men's income to that of their family of origin, that is, their parents' income. Relative income has declined since the late fifties despite the rise in absolute real income.

Source: Appendix Table 3.1.

purchasing power. As can be seen, the line showing the earnings prospects of young men trends generally upward, although it has leveled off in the last few years. This shows that as a general matter the earnings of young men at later dates can buy a greater absolute amount of goods than those of young men at the beginning of the period. But the upward trend in young couples' aspirations, shown by the middle line, is even greater. As a result, the relative income measure, the top line, moves steadily downward, except for a mild interruption in the late 1960s.

Let us reflect on the meaning of the decline in the relative income curve, assuming that the curve successfully captures what it aims to measure. The first point on the curve, that for 1957, shows a value of 73 percent. Now suppose that at that date you were under twenty-five years old and a partner in a typical (or average) American couple considering marriage and children. If you and your partner aspired to a standard of living that (whether or not you were aware of it) corresponded roughly to the income level in the man's family of origin when he was growing up, the man's prospective income could in itself support 73 percent of that desired standard. Contrast this with the situation of another typical young couple in 1978. According to the chart, the man's prospective income, although absolutely greater, would support only 54 percent of the level desired by the couple. Thus, young adults today find themselves under considerably greater economic stress than those in the late 1950s. The generally downward drift of the curve implies that throughout most of the period the situation of young couples deteriorated fairly steadily.

Taken literally, the relative income measure implies that a couple aspires only to the actual income level in the husband's family of origin. One might suppose that a young couple typically aspires to more than their parents had. If this were so, the denominator of the relative income measure should be raised by what one might call an "escalation factor" and the quotient correspondingly decreased.

Such an adjustment would be reasonable, but is unnecessary for now because our interest here is in the way relative income changes over time, that is, in whether or not young adults in the seventies felt under greater economic stress than their counterparts did in the late fifties. It is reasonable to assume that young couples always aspire to more than their parents had—in other words, that the escalation factor applies to values for all dates. An adjustment of this sort would shift the curve downward at every point in time. Thus, one could still conclude that young adults in the seventies were under considerably more economic pressure than those in the late fifties. More generally, an adjustment to the aspiration level of income—for whatever reason and whether upward or downward—would not alter our conclusions about the change over time as long as the adjustment were fairly uniform throughout the period.

Relative Income before 1957. As we push further back in time, the data become increasingly scarce, making the going even tougher. But the greater contrast in the pre- and post-World War II labor markets makes the need for precision less urgent. A young couple's ability to support their aspirations almost certainly must have improved greatly between the years preceding World War II (when about one person in six was out of work) and those after (when those unemployed were fewer than one in twenty). We also know that in 1941 only 60 percent of men in their early twenties earned any income at all, whereas almost 100 percent of their fathers at that time earned incomes. By the late 1940s, this differential between young and old had narrowed sharply, as the proportion of younger men earning income rose to 91 percent.[8]

In the following section, I use, in the absence of data on relative income before 1957, a measure of the comparative unemployment of fathers and sons. The idea behind the measure is this: if young men after World War II experienced better labor market conditions than had their fathers over the course of their working lives, while those before World War II experienced worse conditions compared with their fathers, then young men after World War II were in a better position to support a young couple's aspirations. To measure labor market conditions for fathers and sons, I use the nationwide unemployment rate when both were in the labor market. Hence, the measure is, in effect, one of relative unemployment.

Relative Income, Marriage, and Childbearing. We will now test the relative income hypothesis. The expectation is that if the relative income of young couples improves, marriage and childbearing will be encouraged; if it deteriorates, they will be discouraged.

The swing in the population's average rate of childbearing since World War II—the baby boom and bust—results from two underlying developments. In part, it reflects a corresponding swing in marriage patterns. For example, in 1940 about 10 percent of teenage girls over fourteen were married. After World War II, this climbed to a peak in the late 1950s of nearly 15 percent and since then has dropped, reaching its present low of around 9 percent.[9] More important than the change in marriage patterns in shaping the overall rate of childbearing, however, has been a swing in the average number of children born to married couples (marital fertility), which also reached its peak in the 1950s, and has since dropped sharply.

To test the relative income hypothesis, a measure of childbearing is used, the "total fertility rate," * that reflects the influence of changes in both marriage behavior and marital fertility. Changes in the total fertility rate are dominated by the

* The total fertility rate in any year is the total number of children that a hypothetical woman would have borne if she had gone through her reproductive life having children at the average rate of childbearing actually prevailing in that year at each age from fifteen to forty-four.

childbearing behavior of younger women—those under thirty —who bear more than three fourths of the children.

Figure 3.2 shows the relative income and relative employment measures just discussed, together with the total fertility rate. The similarity between our measures of economic stress and the childbearing of young couples is obvious. The relative economic position of young adults peaks in the late 1950s, similarly to childbearing. When a couple's outlook was exceptionally good, as in the late 1950s, marriage and childbearing were accelerated, and the rate of childbearing was pushed up. When it was comparatively poor, as before World War II and in the late 1970s, marriage was deferred, and childbearing within marriage postponed, depressing the total fertility rate.

Relative Numbers and Fertility. We can try a second test of our hypothesis. In the last chapter, we saw that since 1940 the economic experience of younger men relative to older has been largely a reflection of changes in generation size that affected their relative numbers in the labor market. While the relative income measure used here is not exactly the same as the measures used in chapter 2, nevertheless the movement of the relative income measure is largely shaped by that in the relative number of young men. This suggests the possibility of comparing childbearing directly with a measure of relative numbers.

This comparison is made in figure 3.3, where the measure of relative numbers is the ratio of the male population between ages thirty and sixty-four to that aged fifteen to twenty-nine, the reciprocal of the measure shown in figure 2.1. The rise in this measure to 1955–60 indicates a growing scarcity of younger men relative to older; the decline thereafter, a growing abundance. Note how the fertility curve again is consistent with the hypothesis. Before 1955–60, an increasing scarcity of young men and the consequent improvement in their life chances are accompanied by a rise in fertility; after 1955–60,

FIGURE 3.2

Relative Income and Childbearing

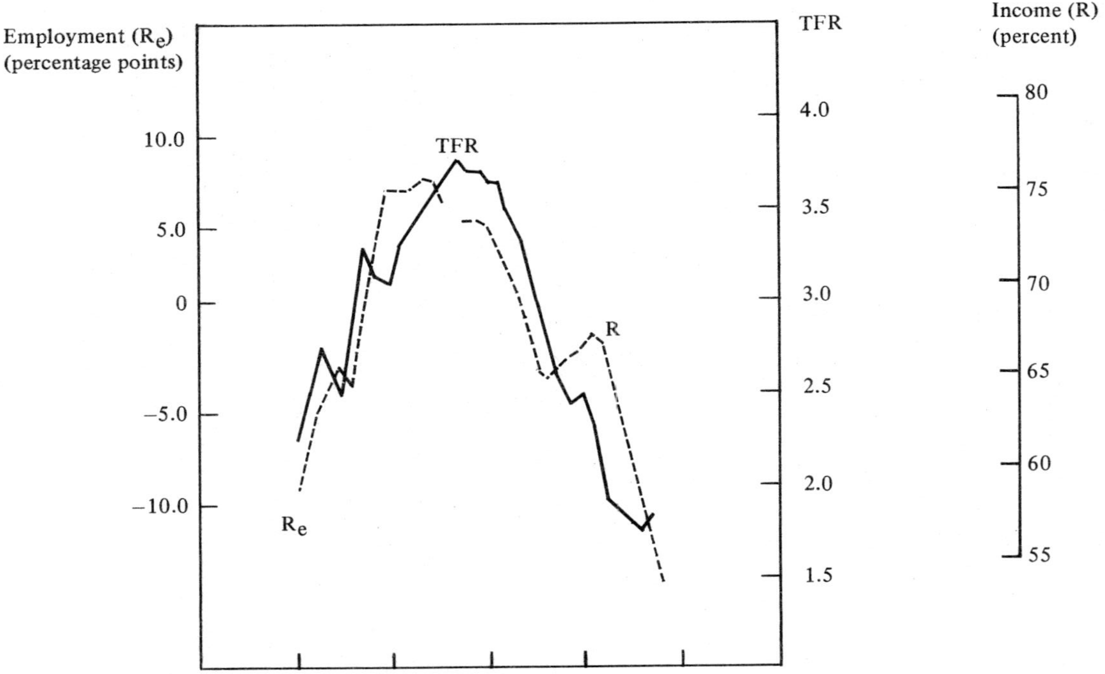

Fertility (TFR) shows a swing since 1940 like that in relative income (R and R_e). (R_e is an approximation to relative income based on employment data.)

Source: Appendix Tables 3.1 and 3.2.

Marriage and Childbearing

FIGURE 3.3

Relative Numbers and Childbearing

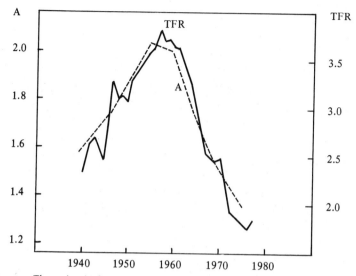

The swing in fertility (TFR) also looks like that in the relative numbers of older to younger adult men (A).
Source: Appendix Table 3.2.

an increasing abundance of young men is paralleled by a decline in fertility. Both figures 3.2 and 3.3 point to the same conclusion: the evidence supports the relative income hypothesis, both for the baby boom and for the baby bust.

Differences among Population Groups. Although both the rise and the fall of fertility were widespread throughout the population, occurring among different racial, ethnic, and socioeconomic groups, there were, of course, differences among population groups.[10] Unfortunately, sufficient data are not available to test whether or not these differences were due to corresponding variations in relative income. An unusual study that traced the experience of a cohort raised in Oakland, California, in the 1930s depression, however, did address this question. This study found that the members of this cohort with the greatest fertility gain also were those with the greatest gain in relative economic standing.[11] Here is one more bit of evidence for the relative income hypothesis.

Variations in Childbearing by Age. One problem remains to be dealt with—an aspect of childbearing behavior in the post-1940 period that would not be explained by the theory so far presented. The emphasis here has been on the childbearing behavior of young adults—those under age thirty. Yet if one compares the fertility change since 1940 of women in their thirties with that for women in their twenties, there is a noticeable similarity. The childbearing rates for both age groups rise and fall with the baby boom and bust, virtually in unison.[12] This synchronism would not be predicted by the theory, and the question is how to explain it?

To do this, we need to add a stipulation about the nature of American family size preference; namely, a notable consensus exists on two-, three-, or four-child families. This is shown by surveys of American women conducted between 1936 and 1972 on their ideal family size—throughout this period, the proportion favoring two to four children is always 85 percent or more.[13] This characteristic of family size preference limits the range within which relative income typically functions. Those persons who in their twenties are in a favorable income situation tend to embark on their childbearing careers early and then in their thirties to cut back as they approach three- to four-child families. Conversely, those in an unfavorable relative income situation in their twenties tend to defer their childbearing careers and then subsequently to make up for this deferment by seeking to reach the desired minimum of two children. We are experiencing the latter development now, as some of the young women who in the seventies were responsible for the baby bust approach or pass the age of thirty and start belatedly to build the minimum desired two-child family. We saw the former development working in the sixties, when the young women who were responsible for the baby boom of the fifties, started to cut back on childbearing as they entered their thirties and their family size approached the upper end of the two- to four-child range. And to go back one more step, the current situation was replicated in the late forties and fifties, as women then in their thirties sought to make up for the lost childbearing of the World War II period.

This compensating behavior produces a swing in the fertility rate for women in their thirties like that for women in their twenties. Thus, in the 1950s, the high fertility rate of young women in their twenties, due to a favorable relative income situation for young adults, was accompanied by a high rate for women in their thirties who were making up for the shortfall of the previous decade. In the 1960s, the declining fertility rate of women in their twenties, due to a deteriorating relative income situation of young adults, was accompanied by a declining fertility rate for women in their thirties who were cutting back as a result of their high fertility a decade earlier. The result has produced a fairly synchronous movement in the fertility rates of women in their twenties and thirties, reflecting the operation of the relative income-fertility relationship within the desired two- to four-child range.

Other Explanations

The relative income hypothesis sounds plausible and fits the data. But there are other persuasive theories of recent fertility behavior. Do they work out as well?

It is impossible to deal with all of the competing hypotheses, but I shall try to touch here and in chapter 4 on the more prominent ones. In evaluating them, the test applied is the same as that for the relative income hypothesis—that the theory fit both the baby boom and the baby bust periods and be consistent with the evidence. (You will recall that I am arguing that my theory has been operative only since 1940.) Even this seemingly simple requirement turns out to be a stringent one, however. Although numerous candidates vie for precedence as *the* theory of the baby bust, there is largely silence on the baby boom. This inability to explain the baby boom is at least admitted by some of the more forthright scholars. Thus, Westoff, whose explanation of the baby bust was quoted earlier, states: "We are beginning to understand the baby boom demographically, although not the underlying causes. . . . What social and economic forces generated and sustained the changes in marriage and childbearing is not at all clear." [14]

Some Nonexplanations. Let me dispose first of some mistaken notions. One is that the baby boom was the result of deferring childbearing during World War II, with the subsequent bust being a return to the prewar trend. As mentioned briefly earlier, those in the 1950s who were principally responsible for the baby boom—young adults in their twenties— were in their teens or preteens during World War II. The reproductive careers of these people were not interrupted by World War II, and their behavior in the fifties clearly cannot be attributed to the war.

Let me forestall, also, a mistaken interpretation of the present argument. Since my theory stresses the relative size of different age groups, readers acquainted with demography may think that I am presenting the demographer's standard "age structure" explanation of fertility. The usual argument is a simple one: fertility is concentrated among those in the reproductive ages. If the share of this group in the total population rises, then, other things being equal, the crude birth rate, which is the ratio of births to the *total* population, will rise. In like manner, a fall in the share in the total population of those in the reproductive ages will depress the birth rate of the population as a whole.

Marriage and Childbearing

Note that my argument is just the opposite. My theory has the population of young adults and the birth rate moving inversely—when the proportion of young adults rises, the birth rate falls. And in fact, this is what has been observed since 1940. In technical terms, my theory concerns the effect of age structure on the "age-specific" fertility of young adults— whether there are more or fewer births, say, per 1,000 females aged twenty to twenty-nine. The standard age structure effect of demography—the "age-composition" effect—deals with the changing share of young adults in the population, assuming their fertility rate is unchanged. Both of these effects—the age specific and the age composition—have operated in the post-1940 period, and in opposite directions, although the one I am concerned with, the age-specific effect, has been the dominant one.

What About the Pill? In the history of American contraceptive technology, June 1960 marks an epochal event. In that month, the long-sought-for oral birth control pill was authorized for use. Since then, use of the pill has grown rapidly. Moreover, in the late 1960s, the pill was followed by the intrauterine device (IUD) and by a widespread liberalization of abortion laws. To many writers, these developments—especially the pill—are the key to the baby bust.[15]

As plausible as this "pill hypothesis" seems, there are a number of reasons to question it. Well before its introduction, most young American families were using contraception. Population surveys in the 1950s show this to be the case.[16] But it could hardly have been otherwise—how else could the early nineteenth-century American birth rate of 50 per 1,000 population—higher than in many less developed countries today— have been more than halved by World War II? The pill was largely a substitution of a new method for old. For some households—especially Catholic households, many of whom previously relied on the rhythm method—the pill may have been a more effective means of preventing conception. But experience shows that if a couple wants to limit fertility they do, and so-called inefficient methods become effective. A valuable study by Jeanne Clare Ridley reveals that in the 1930s the primary contraceptive methods used in the United States to achieve the then-record low fertility were the so-called inefficient condom, withdrawal, and douche.[17] Currently, France and Belgium have childbearing rates like that in the United States, but at least two out of three couples in those countries control their fertility by withdrawal or rhythm.[18]

Moreover, the pill did not accelerate the spread of contraception—the growth in the use of contraceptives was slower in the five years after its introduction than in the five years before.[19] Nor could the pill have been responsible for the early stages of the downturn in fertility that started in the late fifties. According to two leading authorities on the pill, Norman B. Ryder and Charles F. Westoff, "the decline in total fertility [at that time] is well in advance of the increase in pill use." [20] Finally, the pill hypothesis leaves unexplained the pre-1960 baby boom. The post-World War II years usually are not seen as a retrogressive period in the ability of Americans to control their fertility. On the contrary, World War II assured that many more young Americans than ever before were systematically educated in techniques of fertility control (chiefly the condom) as part of their indoctrination in the armed forces. Yet the postwar fertility rate soared despite more universal knowledge on how to prevent conception.

A seeming exception to this view appears in a recent study by Norman Ryder that attributes the 1950s baby boom to "sloppier contraception" and rejects the notion that couples in this period wanted more children.[21] According to this argument, the baby boom and subsequent bust were a product, not of any change in family-size desires, but simply of the efficiency with which couples controlled their fertility. This argument is contradicted, however, by surveys of family-size desires, which show respondents in the 1950s baby boom

Marriage and Childbearing

period favoring larger families than those in the 1930s and 1970s.[22] More fundamentally, the explanation leaves unanswered the question of why couples were more careful in their contraceptive practice in the 1930s (and 1970s) than they were in the 1950s. When Ryder was asked about this, he answered: "[A]n unwanted pregnancy in the 1930s was a far more serious matter because of poor economic conditions and prospects. In the affluent 1950s, it seems couples were less vigilant, in their contraceptive practices." [23] In saying this, the author is in effect adopting the present view that variations in contraceptive practice are not the real cause of fertility swings but, like those swings, are a response to economic pressure on couples. In my view, an improvement in relative income relaxes the economic pressure on couples and encourages them to have more children within the desired two- to four-child range. For some couples, this may be a matter simply of reduced "vigilance" in contraceptive practice; for others, of feeling easier about having that other child that they really wanted.

The possibility is not ruled out, however, that the pill and other changes in contraception availability since 1960 may have had some "add-on" effect in reducing fertility. But the spread of these new techniques, rather than being the principal cause of the decline in fertility, was, like the decline itself, more likely an effect of growing economic pressures felt by young adults. Thus, as the relative income position of young households deteriorated and as they felt more hesitant to have children, they turned eagerly to the new techniques available. In other words, increasing economic stress induced *both* the rapid spread of the new birth control methods and the decreased childbearing.

Female Education. Generally, more educated women have fewer children than less educated. Based on this observation—which compares women of different educations at one point in time—it is a short step to a hypothesis about changes over time, namely, that the general advance in educating women is the cause of their declining fertility. Hence, a popular explanation of the baby bust is the progress of female education.

Today's young women are indeed better educated than yesterday's. In this century, however, the progress of education has been always upward, although at an uneven pace, while fertility has waxed and waned. In 1977, women in their late twenties had completed, on the average, 13.0 years of school, that is, the equivalent of one year of college. This was an increase of 0.7 years over the average schooling of their counterparts in 1960. But 1960 women too were better educated than their predecessors—enjoying an extra 2.2 years of schooling over young women who were in their late twenties in 1940. If the 0.7 increment in young women's schooling between 1960 and 1977 is presumed to account for the baby bust, how can one explain the fact that a 2.2 increment from 1940 to 1960 was accompanied by a baby boom? The "advancing education" hypothesis clearly does not stand up to the test of consistency between the baby boom and baby bust periods.

The "Competitive Disadvantage" of Women. A more sophisticated hypothesis relates to special post-World War II conditions. The preference given to returning war veterans by colleges, it is argued, gave men an even greater economic advantage over women than they had before. The relative competitive disadvantages suffered by women in the labor force encouraged them to marry earlier and to begin childbearing without any thought of careers.[24]

The basic difficulty with this view is that it assumes a competition that barely exists. It is a well-publicized fact that women are concentrated in remarkably few occupations—"chiefly into 20 of the 480 job categories listed by the Labor Department."[25] These occupations are widely regarded as "typically female," and competition with men is minimal.

Moreover, suppose we look at the occupations in which college-educated women are concentrated, teaching and nursing, because the hypothesis applies especially to persons with college educations. For these occupations, no hint exists that the preferences given to returning war veterans by colleges were hurting women's chances in the 1950s. In describing the post-World War II market in these occupations, the Department of Labor's *Occupational Outlook* repeatedly uses such terms as "excellent immediate employment opportunities," "serious shortage," and "unprecedented demand." [26] Anyone with first-hand knowledge of the 1950s will recall how elementary and secondary schools were straining to find and hold qualified teachers. The evidence simply does not support the notion that in the 1950s well-educated young women turned to earlier marriage and childbearing because of inadequate job opportunities caused by male competition.

Women's Work. The recent decline in fertility has been accompanied by a notable increase in young wives' working outside the home. Some see the decline as due to new opportunities that are attractive alternatives to motherhood; others see both decreased fertility and increased female labor force participation as reflections of changes in sex-role attitudes and in the status of women. In my view, neither of these hypotheses is correct. But recent changes in women's work, attitudes, and status are of such importance that they deserve attention in themselves, regardless of their relation to fertility. Hence, we turn to the next chapter.

4

Women's Work

LIKE the inverse action of the two pans on a chemist's scale, the plunging American birth rate of recent years has been counterbalanced by a striking rise in young women working outside the home. In 1960, about two out of five women in the prime childbearing years (twenty to thirty-four) were in the labor market; by 1976, the proportion was three in five. This sharp increase in the labor force participation of young women has been taken by some as a mark of the emerging modern woman—well educated, career oriented, financially independent—a woman freed from the wheel of marriage and childbearing.

This view is largely an illusion, as this chapter will show. True, some women presently in the labor market do fit this picture, and their number is growing. But their proportion of the total is small; the large majority are doing much the same type of work for the same type of pay as in the past, that is, they are in traditionally female jobs and earn considerably less than men. The unusually rapid rise since 1960 in young women working outside the home is chiefly due to the decline in the relative income of young couples as generation size has increased. This interpretation holds up to the test put forward in chapter 3; that is, it is consistent with the experience of the two decades before 1960, when young women's labor force participation stopped rising as relative income rose. Moreover, it is supported, as we shall see, by the changing patterns of work for *older* women, which contrast noticeably with those for younger women before and after 1960.

What Causes What?

Let us first clarify some tricky issues of cause and effect. The contrasting changes in young women's fertility and their work outside the home could be because:

1. Women's work determines their fertility;
2. Women's fertility determines their work;
3. A "third force" determines both fertility and work; or
4. Pure coincidence.

Let us set aside the last possibility as highly unlikely because for most women decisions about childbearing and work outside the home are not likely to be made independently of each other. As an example of the first—causation running from work to fertility—suppose that hiring practices that discriminate against women were suddenly reduced. The resulting rise in employment opportunities might encourage women to work outside the home and to have fewer children. In this case, a change in the factors determining women's work would lead to a reduction in their fertility.

An illustration of the second possibility—causation running from fertility to work—would be a breakthrough in fertility regulation, such as the "morning-after" pill. A reduction in unwanted babies due to the new technique would free women for greater employment outside the home. In this case, a change in a factor determining women's fertility would lead to an increase in their labor force participation.

The "third force" possibility is illustrated by the view in the opening paragraph in this chapter, what one might call the "new woman" hypothesis. If women's attitudes about their roles at home and in the labor market were to change fundamentally—in favor of careers and against families—then fertility and labor force participation would move inversely to each other. But this would occur not because one was causing the other but because both were simultaneously responding to a third force.

I think hypotheses like the above fail to pinpoint the motive force behind recent changes in young women's work and fertility. But my theory shares with the "new woman" idea a third force view of fertility and labor force participation. Thus, I believe that recent developments concerning both fertility and work are a response to a common cause—decreased relative income of young adults resulting from increased generation size. As mentioned, I also shall argue that a third effect stems ultimately from this same cause—the changing work force behavior of *older* women.

Some Facts on Women's Work

First, some of the distinctive features of women's work outside the home as it has evolved in this century must be clarified. As noted at the end of chapter 3, women workers are largely concentrated in a small number of occupations, to repeat, "chiefly into 20 of the 480 job categories listed by the Labor Department." [1] These occupations, which are widely regarded as typically female, include jobs that involve activities similar to those that women have traditionally done within the home: personal service jobs, such as waitressing and hairdressing, as

Women's Work

well as so called nurturant jobs, such as teaching, social work, and nursing. They also include jobs that have been created because of the vast growth in clerical positions as business has grown—secretary, typist, bookkeeper, file clerk, receptionist, telephone operator, and so on. Others are associated with the rise of urban service industries—sales clerk, cashier, bank teller, and proprietor of small shops, such as gift boutiques.

Because of the high degree of occupational segregation, competition between men and women for most jobs is quite limited, other than at the very early working ages, as, for example, in the case of fast-food restaurant employees. Generally, we have what is sometimes called a "dual labor market"—separate markets for women and men.[2] Moreover, the two labor markets differ in a significant way. In chapter 2, it was shown that competition between younger and older men is quite limited because they are at different stages of the career ladder. This is not true, however, in the women's labor market. Because of high turnover in the women's labor force older and younger women are not at different career stages; usually, they compete for the same jobs. In this competition, however, younger women often have an edge. Many female jobs involve contact with the public, and in our society, where "young is beautiful" (especially regarding women), employers are likely to prefer younger over older women. In addition, because educational levels tend to advance from one generation to the next, younger women are better educated than older, and this too gives them a competitive advantage.

The earnings of women are considerably less than those of men—"about three-fifths . . . even when both work full-time at year-round jobs."[3] This differential is not so much a matter of not getting "equal pay for equal work" because, in fact, most men and women are not doing the same work. Rather, it reflects the relatively low pay rates in the jobs typically filled by women.

Finally, as a recent government publication notes, "women, especially married women, move in and out of the labor force more frequently than men, and . . . have considerably fewer years of continuous service with the same employer." [4] This is, of course, partly a reflection of the responsibilities of women in the home, particularly with regard to childbearing and childrearing.

Are Things Different Today? To those who favor the "new woman" view, the rapid increase in work outside the home among young wives reflects the improving status of women in terms of type of job, wages, and education. But the facts run counter to this view—the jobs that young women hold are much the same as ever, and compared with men, their earnings and education have not improved; if anything, they have deteriorated.

I have noted in chapter 1 that women in their early twenties in 1975 aspired to the usual female jobs. If one examines the jobs actually held by these young women in that year the picture is the same, except that, as one would expect, they actually worked at lower status jobs than those to which they aspired.[5]

Have women been moving into higher status "men's jobs?" Some signs indicate a modest change in a few occupational categories: physician-osteopaths (the proportion of females employed rose from 8.9 to 12.8 percent between 1970 and 1976) and lawyers-judges (from 4.7 to 9.2 percent); in others, such as engineers (1.6 to 1.8 percent) and technicians other than medical and dental (down from 14.5 to 13.6 percent), the change was negligible or even negative.[6] In business administration, the Conference Board recently published a survey of several hundred companies that was aimed at finding out what progress women had made in the five years since the beginning of vigorous affirmative action programs in 1970. The results showed only small improvement—in industries where the proportion of female workers was below the economywide average (e.g., heavy manufacturing), the percentage of women among managers and administrators rose from only 5 to 6 percent; in industries where their employment was above average (e.g., retail trade), the proportion rose from 19 to 23 percent.[7]

Regarding wages, the story is again the same—in 1977 among full-time full-year workers, women earned 58.5 percent as much as men—slightly below the long-term average.[8]

Therefore, any recent improvement in the job status of young women has been at best small, and cannot account for the rapid rise in their labor force participation. For most women, their work and their aspirations are not much different from the past. Little evidence on women's work today supports the "new woman" view.

This applies also to women's education. Although it is claimed, as proof of the changing status of women, that "educational differences between the sexes have greatly diminished," [9] the opposite is true. Shown below is the trend since 1940 in the median years of school completed by men and by women aged twenty-five to twenty-nine: [10]

	1940	1950	1960	1970	1977
Males	10.1	12.0	12.3	12.6	13.0
Females	10.5	12.1	12.3	12.5	12.8
Excess of males over females	−0.4	−0.1	0	0.1	0.2

In the 1940s and 1950s, young women enjoyed an educational advantage over young men; today, young men enjoy a slight advantage over young women. As a result of recent college enrollment, a slight shift back in favor of women will likely occur during the next decade, although hardly enough to justify the view that, compared with men, the educational status of women has noticeably improved.

Younger Women's Labor Force Rates

Thus far, the evidence on young women's work and their education largely belies the "new woman" notion. Moreover, no indication exists that affirmative action programs have had a sizable effect on women's work. How, then, can one explain the rapid influx of young women into the labor market?

Recent Experience in Historical Perspective. To help answer this question, let us quickly review the longer term trend in young women's work. As I have said, a valid explanation of recent events must fit the facts of earlier experience as well, and the facts for the two decades before 1960 are in striking contrast with those after. The upper half of figure 4.1 shows the trend in young women's labor force participation in this century and the expected rapid rise since the early 1960s, culminating in the recent all-time highs of around 60 percent. Set against the longer term trend, however, the recent rise is not quite so spectacular. Before 1940, the trend of young women's labor force participation was also steadily upward. True, the rate of advance was slower than that since 1960, but it was nevertheless substantial.

My interest, however, is not in the long-term uptrend as such, which I take as a basic, continuing phenomenon, but in the noticeable deviations that emerge after 1940. Looking at the figure, one sees that the acceleration in young women's work in the two decades after 1960 is balanced by zero growth in the two decades before. What needs to be explained, therefore, is not only the recent accelerated rise but also the puz-

FIGURE 4.1

The Trend in
Women's Work Outside the Home

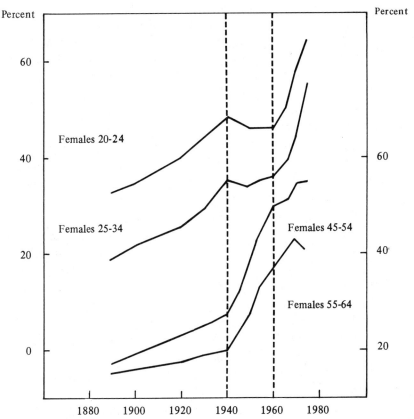

For each age group, the curve shows the percentage of women in the labor force from 1890 to 1975. Before 1940, labor force participation trended upward for both younger and older women. After 1940, younger and older women show sharply contrasting movements.

Source: Appendix Table 4.1.

zling lapse in the years between 1940 and 1960. How can one explain this startling difference?

The Relative Income Explanation Again. My answer is simple and builds on the explanation in chapter 3 of women's childbearing behavior.[11] Consider the situation of a young man and woman who are contemplating marriage. Although both are working full time, their relative income situation is poor, in the sense that both do not make as much as they feel they need to live the way they would like to. What can they do to improve their economic situation? For each, possibilities exist, such as seeking extra training to improve their prospects or moonlighting, that is, taking a second job. But the most obvious adjustment over the long run is a reallocation of the woman's work time from the home to the labor market. For couples who expect to start having children soon after marriage, this means postponing marriage so that the woman can work a longer time. For married couples, it means putting off childbearing. For those who have children, it means trying to combine childbearing and childrearing with work outside the home. Or it may mean having fewer children than were desired so that the wife can return to the labor market sooner than planned. In each of these cases, the solution to the economic pressures is increased labor force participation for the woman. This option is not available to the man because the couple is already planning on his continuous employment throughout the family-forming years.

The link between these options and generation size is obvious. In the 1950s, young workers were in unusually short supply—this was the "scarcity generation" born in the 1930s. For both men and women, job opportunities at good wages were plentiful and unemployment rates relatively low. It was relatively easy to earn enough to satisfy life-style aspirations, in sharp contrast to the pre-World War II situation, when labor demand was low, jobs hard to find, and young persons had a hard time "making ends meet." The favorable employment and earnings experience of young adults in the fifties encouraged earlier marriage and childbearing in two ways. First, many were fairly rapidly able to accumulate a "nest egg" or to acquire those goods needed to establish a home. Second, couples' confidence was increased about the husband's ability to support a household and the wife's being able to give up her job to start a family. Thus, in the 1950s, the high relative income of couples due to their small generation size led to a sharp departure from pre-World War II trends.

In the 1970s, large generation size produced the opposite effect. Relative to their life-style aspirations, young men and women today find it difficult to get by. As shown by the crude relative income measure of chapter 3, young men's ability to supply the couples' needs has declined by close to one-third (see figure 3.1). Since women's economic experience is even more adversely affected by increased generation size, their ability to contribute to the couple's desired life-style has probably declined even more. Hence, young persons are under much greater economic pressure. We have seen some response to this on the part of men, such as increased moonlighting. But the more pervasive response has been on the female side through much more rapidly increased labor force participation.

How successful, compared with young adults in the fifties, have the young adults of the seventies been in realizing their economic aspirations? If, for example, we were to compare a couple whose partners are around age 30 today with a similar couple twenty years ago, would we find that today's couple has realized its life-style aspirations as well? Unfortunately, as indicated in chapter 3, very little has been done to measure economic aspirations; hence, no precise answer is possible. My guess, however, is that today's couple would not show up badly; that is, because of the shift from childbearing to work outside the home, today's young adults have supplemented their income and have achieved, reasonably well, their material aspirations. The difference between today's couples and

FAMILY

those of the 1950s is that the latter were able to realize their economic aspirations along with larger families and less work outside the home for women.*

Older Women's Labor Market Behavior

Relative income as an explanation of recent experience finds further support in the working behavior of older women, who have been lost in the excitement over events in labor force participation of younger women. Only two decades ago, the startling post-World War II upsurge in labor force participation among *older* women was the subject challenging explanation. As this movement has abated in recent years, so too has interest in it. Developments among younger women are now in the limelight and are frequently discussed as though divorced from those among older women. I shall argue, however, that work force changes for older women in the postwar period have been intimately related to those for younger women.

Let us again start with the facts. As I have mentioned, older women hold the same kinds of job as do younger women. What are not the same are the changes in the proportion of each working outside the home. As shown in figure 4.1, before 1940 there was an uptrend in the proportion of older women working outside the home, but the rate of increase was much slower than that for younger women; by 1940, the proportion of women over forty-five working outside the home was still much less than that of younger women. Between 1940 and 1960, how-

* A corollary of the above is that, in any measure of economic pressures that a young couple feels, *actual* (or "family") income relative to economic aspirations is likely to be misleading: their actual income already reflects the adjustments they have made to economic pressure—such as moonlighting or increased work outside the home for the woman.

70

ever, a sharp deviation from the trend occurred, for both older and younger women. For older women, however, the deviation was the opposite of that for younger women—a marked acceleration in the growth of labor force participation, not a leveling off. The previous gap between labor force participation of younger and older women was therefore eliminated—by 1960, work outside the home was as common among older as among younger women. Then in the 1960s, the pattern of change reversed. As work outside the home resumed its upward trend among younger women at a growth rate above the long-term average, that for older women slowed down, falling below the pace of the preceding two decades. By the mid-seventies, as in 1940, work outside the home was once again more common among younger than among older women. Clearly, the problem is to explain why, in light of their similar trends before World War II, the recent period has been marked by opposing movements for the two age groups.

The Effect of World War II. One explanation can be quickly dismissed. Occasionally, the rapid rise in older women's labor force participation after World War II is attributed to the labor shortage during the war, which attracted a number of women into the job market. This experience, it is claimed, encouraged older women to enter the postwar labor market at a higher rate than would have been expected from prewar trends. The problem with this explanation is that it is contradicted by the behavior of younger women, who, during World War II, also enjoyed a disproportionate boost in rates of labor force participation.* However, after the war, labor force participation for younger women fell back to its 1940 level, not even maintaining its prewar rate of growth. If experience during World War II had been such a dominant factor

---

* This brief wartime boost in labor force participation rates is not shown in figure 4.1 in order that long-term trends can stand out more clearly.

in postwar patterns of female work outside the home, the behavior of both younger and older women should reflect that
fact; it does not.

Older Women as Substitutes for Younger. We are back,
then, with the problem of explaining the disparate changes for
older and younger women since World War II. The key to the
answer is provided by the fact that older women and younger
women do essentially the same types of work. In contrast to
men, older women are not on a higher rung of the career
ladder than are younger women. This means that from an employer's point of view a high degree of substitution is possible
between younger and older women.

Consider now the post-1940 labor market for women as a
whole. The demand for female labor was expanding steadily
with the growth of the economy, creating new openings for
secretaries, cashiers, sales clerks, and so on. After World
War II, because the relative income situation of young adults
was so good, the normal growth in labor force participation
of younger women was interrupted as more couples opted for
marriage and a family. With young women failing to enter the
labor market at the normal rate, employers turned increasingly
to older women to fill the gap. In Gertrude Bancroft's words:
"Faced with a restricted number of women in the age groups
that they normally favor, employers were forced to turn to
other ages for their labor supply." [12] The result was a disproportionate growth in work outside the home for older
women. It is worth quoting Bancroft's description more fully:

> With the strong demand for labor during most of the 1950's, based
> in part on the defense program, middle-aged and older women free
> to take jobs outside the home were the only substantial source of
> additional workers. In all probability, if they had not been available
> and if traditional attitudes about the suitability of both married and
> middle-aged women for many jobs had not broken down, the high
> levels of employment of the 1950's could never have been achieved. [13]

What about the subsequent turnaround in the sixties and

Women's Work

seventies—the accelerated growth in work outside the home for younger women and the slowdown in growth for older women? The key is the deterioration in young adults' relative income as generation size increased: young women flooded the labor market at above normal rates, taking advantage of the continuing growth in demand for female labor and partly displacing the normal growth in older women's work outside the home.

In addition, the movements for older women also may have reflected their own relative income influence.[14] Before 1960, the relative income situation of *older men* was adverse, and their wives were under greater pressure to work. After 1960, the situation of older men improved, thus reducing the pressure for their wives to work.

Overview. In sum, in the decades before 1940, as the demand for women workers expanded, both younger and older women entered the labor market at increasing rates, although younger women, whom employers usually favored, entered more rapidly than did older. After 1940, although the demand for female labor as a whole continued to expand, the rate of growth in work outside the home for older women tended to move inversely to that for younger women. In the two decades before 1960 (World War II aside), growth in labor force participation among younger women ceased, despite the excellent job opportunities, because young couples' relative income improved sharply as their relative numbers declined; so they married earlier and started childbearing sooner. Because of the jobs left unfilled by younger women, employers turned to older women, and their labor force participation rose at unusually high rates. The break with the pre-1940 pattern of roughly parallel growth in labor force participation for younger and older women was thus caused by unusual opportunities for marrying and starting families among young couples, due to the favorable impact of generation size on young couples' relative income.

After 1960, as opportunities for marriage and childbearing

worsened with the deterioration of young couples' relative income, the labor market situation reversed. Not only did the historic uptrend in younger women's work outside the home resume, but it did so at a more rapid rate as young wives came under increasing pressure to help support family aspirations. The growth in opportunities for older women was thus increasingly curtailed and led to a slowdown in the growth of their labor force participation.

Labor Force Rates and Fertility: Some Other Views

It is time now to return to the relation between young women's childbearing and their labor force participation, a subject put off at the end of the last chapter. Let me take up first some competing hypotheses.

The Growth of Employment Opportunities for Women as the Cause of the Fertility Decline. Two analysts at the Rand Corporation in Santa Monica, California, recently advanced the view that "young women's fertility has been strongly influenced by increasing demand for female labor." [15] If, for example, the market demand expanded more rapidly after 1960 than before, then many more young women would work outside the home rather than have children. The authors claim that "[t]he prolonged economic expansion of the 1960s, with rising wages and job opportunities, induced increasing numbers of women to work outside their homes, and correspondingly, to forgo, or at least delay, having children. . . . After the 1970 recession, real wages resumed their steep rise and women went to work in record numbers instead of having children." [16]

This theory is at variance with what is known about the female labor market in the post-World War II period. As I have said, the overall demand for female labor has depended primarily on the growth of a limited number of occupations in the professional, clerical, sales, and service fields. Compared with the growth of jobs generally, the occupations in which younger women are chiefly employed did not expand more rapidly after 1960 than before.[17] Hence, the demand for young women's labor was not unusually favorable after 1960, and young women were not disproportionately pulled into the labor market.

Descriptions of opportunities in women's occupations in the Labor Department's *Occupational Outlook Handbook* provide corroboration. The handbooks for the 1950s use such terms as "serious shortage" (elementary school teachers, 1951), "excellent prospects" (secretaries, 1951), "many thousands of job opportunities" (sales clerks, 1959). In contrast, those of the seventies frequently describe women's opportunities in more guarded terms such as "favorable" (nurses, 1978–79), "keen competition" (elementary school teachers, 1978–79). Also, if the demand for young women's labor were unusually favorable since 1960, one would find their wages rising compared with men's and their relative unemployment rates falling. But as we saw in chapter 2, according to these measures the relative position of women has deteriorated since 1960. Finally, if the demand for female labor were unusually favorable, all age groups of women would show similar responses in labor force participation because of the high degree to which older and younger women are in the same jobs. In fact, the rise in young women's participation has been accompanied by a slowing or even cessation in that of older women.

Changes in Attitudes as the Cause of the Shift from Childbearing to Work Outside the Home. The other argument that needs to be dealt with—in this case a sociological one—is that changes in young women's work and fertility since 1960 reflect a fundamental shift in their feelings about the relative merits of work and childbearing. Thus, the women's movement and zero population growth ideology are taken as indicative of new attitudes in favor of work outside the home and against childbearing.

While this view may be correct for some women, the evidence on attitudes toward childbearing and work held by the bulk of the population does not bear it out. As we have seen, most young women still expect to have at least two children, and their job aspirations continue to center on traditional women's jobs (see chapter 1). Moreover, it is not easy to believe that the young wife who shoulders both a job as secretary or supermarket cashier and the job of childrearing does so for ideological reasons.

More plausibly, the increased support both for the women's movement and for zero population growth, rather than causing the shift from childbearing to work outside the home, is at least in part an effect of these developments. The reasoning is simple: the decrease in young couples' relative income has led to reduced childbearing and greater work by wives outside the home. Because of this altered family situation, young men and women favor ideological developments consistent with their behavior. This interpretation is consistent with a psychological theory that carries the impressive title "cognitive dissonance." [18] This theory, which is supported by a variety of studies, maintains that when there is a disparity between attitudes and actual behavior individuals tend to bring their expressed attitudes into line with their behavior. A pertinent illustration is provided by a recent study that found more favorable attitudes about women with young children working outside the home. Commenting on this, the authors point out that "the very fact that more women are now working even when their children are young . . . may mean it has become increasingly difficult for them to believe that this activity is harmful for marriage or for children's well-being." [19]

I am not suggesting that the changes in women's work and childbearing explain entirely more favorable attitudes toward the women's movement and zero population growth, or that these attitudes, for whatever reason they occurred, will not be lasting. I am instead arguing that the shift from childbearing to work among young women is due to the economic pressures that young couples were feeling, pressures that created a situation that made them more receptive to ideological developments supportive of their behavior.

Putting It All Together

Let me summarize my view on the causal connection between women's work and fertility, adoption of new contraceptive methods, and expressed changes in attitudes of a seeming pro-work anti-childbearing nature. Under the special conditions existing since 1940, the ability of young couples to support their economic aspirations has depended to an unprecedented degree on their generation size. The sharply declining generation size of young persons reaching the labor market in the two decades before 1960 increased their ability to realize their economic aspirations, that is, their relative income. In turn, this had a major impact on marriage, fertility, and female labor force participation. Among younger women, marriage and childbearing were speeded up and the long-term uptrend in work outside the home interrupted. Among older women, the long-term uptrend in labor force participation was accelerated as they took advantage of job opportunities passed up by younger women. After 1960, as increasing generation size caused the relative income of young couples to decline, childbearing among younger women was reduced or deferred and work outside the home accelerated, sometimes in conjunction with child-rearing. The economic pressures young couples experienced also led them to adopt newly available methods of contraception in order to implement this shift from work to home. The rapid rise in young women's labor force participation led to a corresponding interruption in the uptrend in older women's labor force participation rates, as employers shifted back toward the age group they traditionally preferred. The shift from childbearing toward work on the part of younger women strengthened attitudinal shifts of a pro-work anti-childbearing nature, such as the women's movement and zero population growth. Thus, shifts in generation size that caused major changes in the economic pressures felt by young couples since World War II had a pervasive impact on both attitudes and behavior relating to work and childbearing.

5

Breakdown of the Family?

MY DISCUSSION so far has been in terms of the traditional two-parent family. But questions have been raised about whether or not the family as we have known it is becoming obsolete. The two leading indicators of family breakdown—divorce and illegitimacy—have taken a sharp turn for the worse in the last two decades. Divorce signifies the dissolution of an established family; illegitimacy an "uncompleted family." [1] Because of the rapid rise in divorce rates and illegitimacy, one-parent families have grown markedly since 1960. Today, only two children out of three live with their own, once-married parents. [2]

Despite these developments, the traditional family remains the prevailing form. In the words of two leading experts, Paul C. Glick and Arthur J. Norton, "most Americans still experience some variation of the 'typical' family life cycle of the past; two out of three first marriages taking place today are expected to last 'until death do them part.' " [3] Hence, my emphasis on the traditional family is reasonable.

Moreover, the picture may not be so bad as it seems—recent changes in divorce and illegitimacy may exaggerate the long-term trend. Generation size also has influenced family disorganization. Small generations typically have an unusually low incidence of divorce and illegitimacy; large generations, an unusually high incidence. Generation size, however, is not the whole story: longer term forces, such as the so-called sexual revolution that has led to greater premarital and extramarital intercourse, are clearly at work too.[4] What seems to have happened is that since World War II changes in generation size have significantly modified the long-term trend—slowing it before 1960 and accelerating it thereafter. Because of this, the experience of the last two decades may give a distorted idea of the long-term trend—the traditional family may not be going down the drain quite so fast as some think.

Of course, not all observers despair about the growth in divorce and illegitimacy. To some, the greater frequency of divorce means fewer spouses locked into unhappy marriages. And a woman going it alone with her illegitimate child may be symbolic of her newfound independence. My concern, however, is not with questions of good or bad. In discussing divorce and illegitimacy as aspects of family breakdown or disorganization, I am interested in "why," specifically how they have been connected to generation size.

Divorce and Generation Size

Let us start with divorce. Why should marriages among members of a large generation be more likely to end in divorce than those of a small generation? To answer this, we must recall the discussion in chapter 1 concerning the different roles that husband and wife expect to play in the family and that society expects them to play.

Role Fulfillment and Divorce. Let me start by contrasting two typical couples.[5] The first is from a small generation—the partners are born in the 1930s and reach adulthood in the 1950s. The second is from a large generation—the partners are born in the 1950s and reach adulthood in the 1970s.

Breakdown of the Family?

Because of their similar socialization experiences, both couples share certain expectations as they reach adulthood. The man and woman will both work before marriage. At some point after getting married—not necessarily immediately after—they will start to raise a family. Before that, the wife will have a job most or all of the time, but when children come along, she will probably drop out of the labor market if the couple can "afford" it and will devote herself to raising the children while they are young. At some later date, she will probably return to the paid labor force.

In the eyes of both partners, and society's too, the man's success is critically linked to his work. His job will determine the social status of the couple and how well they achieve their material aspirations. If at mid-career he can point to a nice family living in a good home in a pleasant neighborhood, then he will have a feeling of considerable personal accomplishment.

The woman's sense of personal fulfillment, and society's judgment of her, will, in contrast, depend much more on her success as a parent than as a worker. If she is one of the relatively small percentage of wives who works at a widely respected job, so much the better. But even if her work outside the home is menial, she will be well regarded and will think well of herself if she has been a "good mother." In contrast, the man is typically judged chiefly on the basis of whether or not he is a "good provider."

The difference in the roles means that the considerations that enter into the choice of a partner are somewhat different for the two. Sociologist Willard Waller once observed: "A man, when he marries, chooses a companion and perhaps a helpmate, but a woman chooses a companion and at the same time a standard of living." [6] This is an exaggeration: economic potential enters into the thinking of both prospective partners. However, expectations about economic performance probably play a larger part in the woman's choice than in the man's. The labor market circumstances of men and women reinforce this—much more variability exists in the potential earnings of men than in those of women. Hence, men are less likely than are women to see big differences among potential partners in what they can contribute to the family income.

These, therefore, are the expectations with which the partners in both couples start their adult lives. But how things work out will be quite different, depending on whether they are from a small or a large generation. The couple from the small generation finds it relatively easy to achieve its expectations. Because their relative income is high, they feel secure about their economic prospects and comfortable about marrying and having children. As their prosperity continues, they find it relatively easy to fulfill their expected social roles.

Not so for the two from the large generation. Even before marriage, they feel the pinch of economic pressure as the competition of large numbers raises unemployment among their cohort and reduces relative earnings. They must wait longer to marry, and they are likely to scale down their expectations about how many children they will have and revise upward their views on how much the wife will have to work outside the home. Nevertheless, both partners retain the same notions about their respective roles, and their self-fulfillment depends in large part on their achievement of these roles. They can not easily shed the skins grown in the long process of sex-role socialization. But the economic bind that they are in makes role fulfillment more difficult. The woman cannot easily give up her job to have children. If she does, she feels under pressure to return to the labor market because her husband's income is insufficient for the household's needs. She feels torn between her desire to care for the children and the need to supplement family income. In such circumstances, a wife naturally feels resentment toward her husband for not living up to expectations, although some wives may not feel this way and others may suppress their feelings. As for the husband, he is likely to feel increasingly inadequate. His progress up the career ladder is disappointing, and he is not able to support his family as he had hoped. He may feel guilty because he and his wife would like to have more children, but she has to work.

Of such stuff are marital strains bred. Moreover, these strains are aggravated by the impact of a couple's relative income on their childbearing. Children help to hold a marriage together.[7] They create emotional ties, and a husband or wife will think twice about leaving, not only one's spouse, but also one's children. If as a result of economic pressures a couple's childbearing is reduced or eliminated, one of the important bonds of marriage is correspondingly relaxed. Moreover, with the wife working outside the home, her feelings of financial independence may be increased. Thus, in a large generation, marital strains are increased both directly, through the difficulty spouses have in fulfilling their expected roles, and indirectly, through the reduced number of children they have.

Marital strains, in turn, may breed marital dissolution. Not for all couples, of course, and not necessarily right away. I am suggesting that for marriages among a large generation the frequency of divorce is likely to be higher than that among a small generation. I do not mean to suggest that generation size is the only factor in divorce rates: longer term forces are certainly at work. It is widely thought, for example, that the decreasing role that religion plays in shaping family attitudes has contributed to a long-term rise in divorce rates. Other factors have operated sporadically—an example is the recent liberalization of divorce laws (although some evidence indicates that this has so far had relatively little effect).[8] I believe, however, that generation size modifies this rising trend—when a small generation is in the family-forming ages, the rise in divorce rates is slowed; when a large generation is at this stage, the rise is aggravated.

Some Evidence. This argument is supported in somewhat different ways by two studies. The first, by two demographers at the University of Michigan, Lolagene C. Coombs and Zena

Zumeta, demonstrates the importance of the wife's expectations about the husband's economic performance as a cause of divorce.[9] This study was based on a survey of persons in the Detroit area, and is especially valuable because the partners in the marriages studied—whether they ended up intact or broken —were first surveyed when the marriage was still intact. In effect, the authors were trying to find the family circumstances that could have predicted the likelihood of divorce. Hence, the study avoided a problem that occurs when data are obtained after divorce occurs, namely, that after having been divorced the respondents may rationalize what has already happened.

As one would expect, the authors found that the economic circumstances of the couple were an important factor in divorce. For one thing, husbands in those marriages that ended in divorce had had lower earnings than those in still intact marriages. "Perhaps more important, however . . . [were] the expectations and attitudes of their wives toward their economic situation."[10] A significantly larger proportion of the wives whose marriages ended in divorce considered their income to be inadequate and felt that their husbands were doing less well than others with similar work and education. Note that the feelings expressed here involve a comparison of economic performance with expectations—it is not a matter simply of whether or not income is low but whether or not it is low relative to some standard of what "ought to be." Note also that these feelings were being voiced while the marriages were still intact. The lesson is clear: when the husband is falling short of expectations, that is, when the couple's relative income is low, the probability of divorce is increased.

The second study, by demographer Samuel H. Preston and John McDonald, provides further support for this view.[11] Moreover, and this is of special interest here, it tries to explain changes over time in the frequency of divorce.

The authors initially present some unique estimates of the likelihood of divorce among couples first married in a particu-

Breakdown of the Family?

lar year (or period) (see figure 5.1). For example, they estimate that about 15 percent of the first marriages occurring in
1910–14 eventually ended in divorce; for first marriages occurring at the end of the period they cover, 1955–59, the
corresponding proportion is almost one-third. The latter figure
is based on some guesswork—namely, how many 1955–59 marriages currently intact will end in divorce—but the estimate is
not bad because the probability of divorce declines considerably after the first few years of marriage. They chose not to make

FIGURE 5.1

The Trend in Divorce

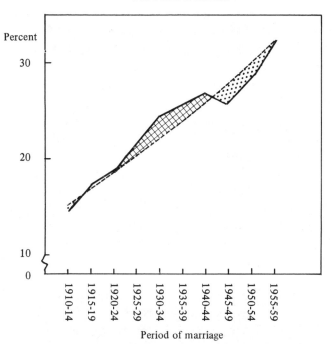

Period of marriage

The solid line shows the percentage of marriages ending in
divorce for couples first *married* in the period shown at the
bottom of the chart; the broken line shows the average trend
during the period.

Source: Appendix Table 5.1.

estimates for couples married more recently than 1955–59 because then the guesswork would have been too great. As we shall
see, however, based on what we already know about divorce rates
among couples married since then, their analysis can be extended
to more recent experience.

As is widely known, divorce has over time tended to become
more frequent. This is shown by the generally upward drift
of the solid line in figure 5.1. Less well recognized, however, is
the fact that this trend has been far from regular—periods
occur in which divorce surges sharply upward and in which it
drops off noticeably. In figure 5.1, this is brought out if one compares the actual percentage divorced (the solid line) with the
"expected" percentage divorced (the broken line), that is, the
percentage that would have been divorced if the uptrend had been
constant. The difference for any given date between the actual
and the "expected" divorce percentage can be thought of as "unexplained"—what one would not have predicted from the general
trend. For years in which this difference is positive, shown by the
crosshatched area in the figure, divorce was unexpectedly high;
for those in which the difference is negative, shown by the speck-
led area, divorce was unexpectedly low.

Note that divorce was unusually high for marriages occurring during the Depression and World War II and was unusually
low for those occurring during the post-World War II period
and through the late 1950s. And although it is not possible now
to set a precise figure, divorce will again be unusually high for
marriages occurring from the mid-sixties through the mid-
seventies. This is a safe forecast because we know that *annual*
divorce rates have been very high in this period, and most divorces occur relatively soon after marriage.

The Preston–McDonald analysis brings out these changes.
What is particularly pertinent here is their explanation of the
unexpectedly high or low rates—the deviations from the smooth
trend shown in figure 5.1. These deviations, the unexplained
divorce percentage at each date, are plotted separately as the

Breakdown of the Family?

bottom curve in figure 5.2. The authors contrast this curve with one for childbearing—the average number of children born to couples marrying in the years shown—the upper curve in figure 5.2. Note how the two curves move in opposite directions —among couples with low fertility, those married in the 1930s, divorce is unusually high. Correspondingly, among those with high fertility, those married at the beginning and end of the period shown, divorce is unusually low. Thus, the parents of the baby boom period are distinctive not only for their high fertility but also for their unexpectedly low divorce rate. We can anticipate that this relationship has persisted into the more recent period—the parents of the baby bust period will be notable not only for their low fertility but also for their unusually high divorce rate.

The inverse movements of the two curves might easily be interpreted as reflecting cause and effect—that low childbearing makes for high divorce, and vice versa—and as I have indicated, some evidence supports this position. But the authors do not take this route. Instead, they hypothesize that divorce may reflect a relative income influence, such as the one I postulated in chapter 3 for childbearing. They test this by comparing the change in a relative income measure that they construct with that in divorce. Let me skip over the details of the comparison and go to their results and their interpretations.[12]

Preston and McDonald find that divorce was unusually high in periods when many couples had earnings that were low compared with the incomes of the families in which they grew up. This was true, for example, of couples married in the early 1930s. On the other hand, divorce was unusually low when most couples' incomes compared favorably with those of their families of origin, as with those married in the late forties and early fifties. Moreover, the analysis suggests that their argument holds as well for more recent marriages. Among couples married in the 1960s and 1970s, relative income (according to the somewhat different measure in chapter 3) has deteriorated,

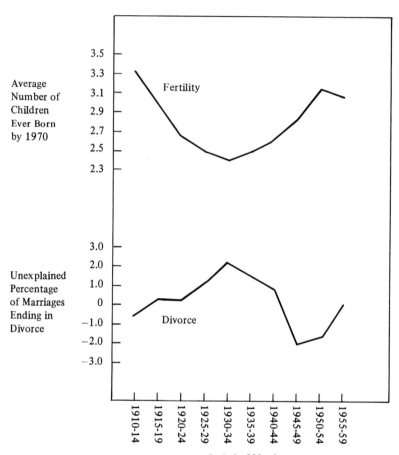

FIGURE 5.2

Fertility and Divorce

Average
Number of
Children
Ever Born
by 1970

3.5
3.3
3.1
2.9
2.7
2.5
2.3

Fertility

Unexplained
Percentage
of Marriages
Ending in
Divorce

3.0
2.0
1.0
0
−1.0
−2.0
−3.0

Divorce

1910-14
1915-19
1920-24
1925-29
1930-34
1935-39
1940-44
1945-49
1950-54
1955-59

Period of Marriage

Lower fertility and more frequent divorce go together. The upper curve shows the average number of children born per woman first married in the period shown; the lower curve, the "unexplained" divorce percentage in figure 5.1, that is, the deviation of the actual divorce percentage from the trend shown by the broken line.

Source: Appendix Table 5.1.

and divorce has been rising rapidly. Income in the family of origin is significant, the authors argue, because it establishes the consumption aspirations of each generation. Hence, when earnings are low relative to aspirations, divorce is more frequent; when earnings are relatively high, divorce is less frequent.

Thus, the Coombs–Zumeta and Preston–McDonald studies demonstrate, in somewhat different ways and on two different sets of data, the point being made here—high relative income makes for a lower probability of divorce; low relative income, a higher probability. In addition, relative income also tends to affect divorce indirectly through its impact on childbearing—high relative income tends to raise childbearing and to lower the probability of divorce; low relative income, to reduce childbearing and to raise the likelihood of divorce. This indirect effect of relative income on divorce reinforces the direct effect.

What does all this have to do with generation size? The connection is obvious. As we have seen in chapter 2 and 3, small generations typically have high relative incomes, thanks to their scarcity. Marital strains are thus reduced and the likelihood of divorce lessened. The opposite is true for large generations. This implies that the long-term uptrend in divorce was reduced in the fifties, when young adults were from a small generation, and has been aggravated in the seventies, when young adults were from a large generation.

Illegitimacy and Generation Size

As I have mentioned, illegitimacy is usually regarded along with divorce as a key indicator of family breakdown, a symptom of "uncompleted families." In this discussion, I shall focus on teenagers. Not only is illegitimacy concentrated among teenagers—in 1975 they accounted for over half of all such births —but it arouses the greatest social concern. My interest, as it was for divorce, is not to discuss all of the causes, but to note how variations in the life chances of young persons caused by generation size have since 1940 left their mark on teenage illegitimacy.

Teenage Sex, Pregnancy, and Marriage: In Theory. The reasoning linking illegitimacy to the relative income of young persons and thus to generation size, can be quickly summarized. I assume that many teenagers who engage in sexual intercourse expect eventually to marry their partners. If pregnancy occurs, then the child is likely to be legitimated by the couple's marrying before the child's birth. Whether or not a conception is in fact legitimated depends partly on the strength of the marriage market. As we have seen in chapter 3, the strength of the marriage market depends, among other things, on relative income —the earnings prospects of young adults relative to their material expectations. Relative income, in turn, varies inversely with generation size.

Consider one imaginary couple who have some vague notion, explicit or implicit, of eventually marrying, the timing depending on a number of things, among them, the couple's capacity to support their material "needs." They are engaging in sexual intercourse, and in time, the woman becomes pregnant. A new factor has entered the marriage picture, one strongly predisposing the couple toward early marriage. But both partners will weigh this factor against others. The young man, in particular, will question whether or not he can support a family. If times are good—that is, if he can meet reasonably well the couple's material aspirations—no serious conflict will arise, and the premarital conception will be legitimated. If times are bad, however, the young man may be confronted with the difficult problem of providing for the household's support. Some couples may resort to abortion, an option increasingly available.

Breakdown of the Family?

Among those couples that do not, some will go ahead with marriage despite the bleak financial outlook. Their problem may be alleviated by the prospect of the wife's working, although the earnings potential of a teenage mother is not likely to be high; moreover, if the labor market for young men is poor, the same is true of that for young women. But inevitably, some couples will break up because of the young man's inability or unwillingness to shoulder the burden of family responsibilities. The result will be illegitimate births that, had times been better, might have been legitimated.

This is clearly not the story behind all illegitimate births, but it is a plausible account of some. I believe that those teenage couples who are confronted with the tension between pregnancy and a difficult relative income situation are numerous enough to have a noticeable impact on illegitimacy. Under these circumstances, if generation size is small and young persons are doing well, marriage will legitimate more premarital conceptions and lower illegitimacy rates; if generation size is large and young adults are doing poorly, the opposite will be true. Thus, legitimation of premarital conceptions by marriage should vary directly with relative income, and illegitimacy should vary inversely.

Teenage Sex, Pregnancy, and Marriage: In Practice. Until recently, only a few scraps of information were available about sexual activities and the legitimation of premarital conceptions among teenagers. This void has been filled by two pioneering national surveys conducted jointly by John F. Kantner and Melvin Zelnik,[13] and by Census Bureau studies by Martin O'Connell and Maurice J. Moore.[14] The Kantner–Zelnik surveys provide documentation for the first part of the reasoning above, namely, that for a number of teenagers premarital sexual activity is linked to marital expectations. In 1971, about one-fifth of unmarried white females between ages fifteen and nineteen reported that they were sexually experienced. Of these women, more than 60 percent had had intercourse with only one partner, many saying that the partner was the man they intended to marry.

Among *all* (not just sexually experienced) white teenage females in the early 1970s, the proportion that by age nineteen had had a premarital conception terminating in a live birth is estimated at about 12 percent.[15] Not all of these conceptions ended up as illegitimate births, because about two-thirds were legitimated by marriage before birth. This figure gives an idea of the importance of the legitimation process that is emphasized here.

Relative Income and Legitimation. Our interest is in whether or not, as predicted, legitimation has varied over recent decades, that is, whether or not rising relative income of young males has been associated with increased legitimation, and vice versa. To answer this, we have the benefit of the recent O'Connell–Moore study,[16] according to which legitimation of premaritally conceived first births reached a post-World War II peak in the late 1950s, when three out of four premarital conceptions were legitimated (see appendix table 5.2). Since then, legitimation has declined to a recent low of less than six conceptions in ten.

Would the observed shift in legitimation have a sizable impact on the illegitimate birth rate? The answer to this is yes. If the proportion of legitimated births dropped from around 75 to less than 60 percent—roughly the decline shown from the late fifties to the late seventies—other things being constant, the illegitimate birth rate would rise by about six-tenths, somewhat more than the actual increase during this period.[17] "Other things" are not constant, of course, but a large share of the recent rise in the illegitimacy rate must reflect the decline in legitimation.

A precise comparison of legitimation with relative income is not possible because no good measure of relative income is

Breakdown of the Family?

available for teenage girls and their partners. But the movement in legitimation is roughly consonant both with relative income changes for the somewhat older group of young adults shown in figure 3.1 and with the view that small generations have relatively low illegitimacy and large generations relatively high.

Breakdown of the Family?

The point of this discussion of divorce and illegitimacy is quickly made. Recent increases have engendered fears for the future of the American family. To the extent that these concerns are based on longer term forces at work—and such forces do exist—they are well founded. The present analysis suggests, however, that the recent upsurge in rates of divorce and illegitimacy is partly due to an increase in generation size. Large generation size means that young persons will have trouble earning enough to support what they view as a family's "needs." In such a situation, if a young unmarried couple engages in sexual intercourse and the woman becomes pregnant, explicit or implicit understandings about eventual marriage are less likely to be fulfilled, and illegitimacy will rise. Among married couples, the economic stress caused by large generation size raises marital strains in a variety of ways. The husband is likely to feel that he is not adequately fulfilling his role as a "good provider." The mother may feel a conflict between her roles as mother and worker—the pressure on her to supplement the family's income by work outside the home may interfere with the care and attention she feels she should give to her children. And with economic pressures leading to lessened childbearing or even to childlessness, spouses are under less constraint to hold a marriage together "for the children's sake." All of this means that the frequency of divorce, like that of illegitimacy, tends to be higher among large generations than among small and that the shift from small to large generations in the two decades before and after 1960 has created an exaggerated notion of the trend in family disorganization. Although changes are taking place in the traditional family, in the long term they may prove to be slower than recent events imply.

III

Society and
Economy

6

Social Disorganization

AS WE HAVE SEEN, some of the symptoms of the recent rise in family disorganization can be attributed to increased generation size. Can any other signs of social deterioration be ascribed to the same cause? The answer is yes. Crime, suicide, and political alienation are more prevalent among young adults from large generations than among those from small. So too are symptoms of psychological stress, such as nervousness and headaches. Moreover, older adults show changes different from those of younger that seem related to differences in generation size. In this chapter, I shall connect generation size and these various signs of social disorganization and mental stress.[1] Let me say once again that I do not claim to deal exhaustively with these topics. I think that it can be shown, however, that logical connections exist between generation size and social and psychological conditions and that the evidence supports such relationships.

Generation Size and Mental Stress

Why should the size of one's generation affect the prevalence of mental stress?[2] The answer has already been suggested by the preceding chapters. Most young persons approach adult life with the expectation of getting married, having children, and living a "normal" family life. A couple from a large generation finds it difficult to support the life-style they desire; in other words, their relative income is low. Both partners have a higher probability of being unemployed than those from a small generation. Pay rates are disproportionately low relative to those of older persons and job advancement slow. The partners are consequently hesitant to marry, and if they do, they are likely to feel pressured to supplement family income and uncertain about whether or not they can afford to have a child "yet." If they do have children, the pressure to bring in more money means that they give less attention to the children than they would like and, for that matter, to each other. They have difficulty, in other words, in establishing the kind of life that they would like to live, in terms of combining work outside the home with their marital and parental responsibilities. As a result, marital strains occur. The idea, of course, is not that members of small generations are immune to such strains and pressures—virtually everyone experiences such conditions at one time or another. Such problems, however, are likely to be more prevalent in a large generation.

Psychologically, these pressures are likely to be translated into increased mental stress. The young man or woman who has difficulty getting a "decent" job or securing a desired promotion is likely to feel increasingly frustrated and doubt his or her ability. The reaction may take the form of resentment toward others—to blame, say, one's employer, "politicians," or society generally. Having to put off marrying or childbearing can also be frustrating. For those who do have a family, the problem of conflicting responsibilities arises. Parents wonder if they are spending enough time with children—or if they are, they may feel guilty that they are not working hard enough to get ahead on the job. A damaged self-image may result from failure to come up to society's and one's own expectations.

I do not want to belabor this argument. Humans are re-

markably pliant and survive surprisingly well under many difficult circumstances. But I think that a reasonable case can be
made to show that a large generation is likely to experience a
greater frequency of mental stress than is a small one—that
such feelings as inadequacy, hopelessness, despair, resentment,
bitterness, or rage may be more common. The difference between large and small generations is, of course, one of degree,
not of kind.

Mental stress increases because large generation size weak-
ens the economic underpinning of personal life. As I mentioned earlier, the difficulties of a large generation start early
in life—possibly with feelings of being neglected at home or in
school, failure to make the school team, and so on. The economic difficulties encountered when working age is reached are
just one more confirmation that life is tough, but it is a crucial
one because economic success is so fundamental to the attainment of most aspirations. 205734

Evidence of Mental Stress. Does the evidence support this
reasoning? The answer is yes. In recent decades, a number of
surveys have probed people's feelings, and sufficient sophis-
tication has been achieved in analyzing these data to permit
the drawing of reasonably reliable inferences. One study ad-
mirably suited to the present purpose has recently been done
by Joseph Veroff of the University of Michigan's Survey Research Center.[3] Veroff compared two nearly identical national
surveys that inquired into feelings of well being and distress.
One of these surveys was made in 1957, when younger adults
were benefiting from small generation size, and the other in
1976, when younger adults were suffering from the disadvantages of unusually large generation size. (Veroff was not studying the effects of generation size—it just happens that his dates
work out well for my purpose.)

His results show a marked increase in psychological stress
between the two dates. The percentage of persons in their
twenties reporting that they were "very happy" declined from

99

39 to 31 percent, and the percentage reporting that they worry "a lot" or "always" rose from 32 to 51 percent. The survey also inquired into symptoms of psychological anxiety, such as insomnia, nervousness, headaches, loss of appetite, and nausea. The percentage of young adults very high on these measures of anxiety rose over the two decades from 10 to 16 percent. A clear and consistent picture emerges of greater psychological tension among the large generation of the seventies than among the small generation of the fifties.

Younger versus Older Men. A large generation is subject to disproportionately high mental stress throughout the life cycle, a small generation, to relatively low stress. At any given time, however, large and small generations coexist. Young adults today are members of a large generation, but their parents, who are now mainly in their forties and fifties, are members of a small generation. If the theory about the psychological impact of generation size is correct, then the changes observed among younger adults would not necessarily be the same as those for older adults. And this turns out to be true. A comparison of older adults at the two dates shows that feelings of well being underwent a less negative or a more positive change than did those of younger adults. For example, while the percentage of those who said they were very happy declined between 1957 and 1976 from 39 to 31 percent among twenty-year-olds, it remained constant among forty-year-olds and declined only from 35 to 31 percent among fifty-year-olds (see appendix table 6.1). Relative size has therefore been affecting differentially the degree of psychological stress experienced by different generations.

If only a few surveys of subjective feelings had been made, one might hesitate to draw these conclusions. The question would arise, for example, of whether or not respondents are telling the truth, or if they are, whether or not states of mind may be so variable that inferences about changes over time and differences among population groups are invalidated. However, enough surveys have now been conducted, taking into account these difficulties, that the data problems that do exist do not upset the present conclusions.[4]

Crime and Suicide

If a large generation experiences more psychological stress, it is likely to show behavior symptomatic of such stress. If feelings of inadequacy are more prevalent, mental depression and, at the extreme, suicide will be more frequent. If feelings of resentment and bitterness occur more often, antisocial behavior, such as crime, is likely to be more common. Thus, larger generations are likely to show higher rates of crime and suicide than smaller. These are, of course, extreme types of behavior, confined to a very small proportion of a generation's members. But they are symptomatic of strains that are more widespread. Hence, a look at the evidence on crime and suicide is desirable, not only because of the interest inherent in these subjects as social problems, but also because it provides further confirmation of the psychological impact of generation size.

Crime. "Crime Amidst Plenty: The Paradox of the Sixties" —this is how one scholarly study highlighted the problem of the rise in crime rates that first appeared in the sixties and that continued in the seventies.[5]

At first, the significance of this rise was discounted as a reflection of a simple demographic phenomenon, one different from that stressed here. Younger people are more likely than are older to be involved in criminal activity; hence, a rise in the proportion of younger persons pushes up the average crime rate, even though the frequency of crime does not change among either younger or older persons.

This development has, in fact, contributed in the last two

decades to the rise in the average crime rate. But further analysis shows that an important factor has also been the *growing* frequency of crime among the young. For example, sociologist Charles F. Wellford, studying crime statistics from 1958 to 1969, concludes that "the rate of crime among those aged 15–29 has increased dramatically in the last decade. . . . [T]he phenomenon of youthful violent crime must be given greater attention in our studies of crime causation." [6]

Wellford's conclusion is supported and made contemporaneous by the record of homicide rates, which is generally agreed to be the best single measure of long-term changes in crime (see figure 6.1, top panel).[7] Although these data refer to victims rather than to perpetrators, they can be used here because murder is much more common among younger men than among other groups in the population and because the victims are usually members of the same age group.[8]

Following World War II, deaths due to homicide among men aged fifteen to twenty-four declined slowly to a low in the middle and late fifties. In the early sixties, the homicide rate edged up slightly; then, starting around 1964, it increased dramatically, reaching a peak in the mid-seventies that was well over twice as high as the 1955 low.

This increase in the homicide rate among young people (indeed, in the crime rate generally) shows the impact of generation size. If the rise in the average crime rate for the whole population were merely due to a larger proportion of younger persons in the population, without any increased frequency among the young, then to infer that the rise in crime reflected increased psychological stress among the young would be a mistake. But in fact, the rising crime rate is also due to a higher rate of youthful crime. In the period through the late fifties, when young adults were from the small generation born in the thirties, the homicide rate was declining or was steady. From 1960 onward, when generation size started to surge upward, criminal activity among young men also rose. This can be

FIGURE 6.1

Crime and Suicide Rates

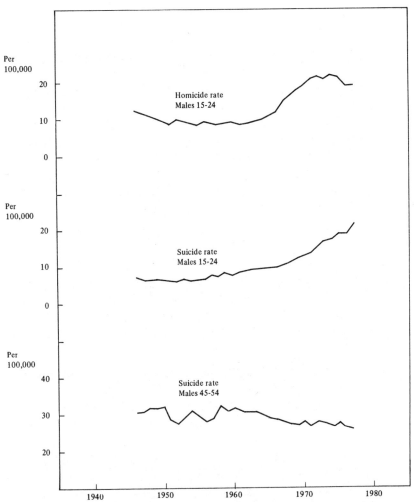

Following a period of slight decline and stability in the post-World War II period, young men's crime rates (as shown by the upper panel) and suicide rates (middle panel) turned sharply upward after 1960, as generation size increased. In contrast, the suicide rate of older men after 1960 (lower panel) did not follow the rising path taken by that for younger men.

Source: Appendix Table 6.2.

ascribed partly to frustrations, resentment, and bitterness from the increased difficulty that young men were having in the job market and in achieving their life-style expectations.

Homicide is considerably more frequent among blacks than among whites; hence, the question arises whether or not the recent rise is due to changes peculiar to the black population. It is not—the rise in homicide rates occurs among both whites and blacks; proportionately, the increase actually is greater for whites.[9]

Suicide. Some people blame society for their failure to live up to expectations and react against it, sometimes criminally. Others blame themselves. At the extreme, the result may be suicide or attempted suicide. If generation size affects the relative income of young adults, their suicide rate would presumably reflect variations in stress associated with generation size. In fact, this is so. The suicide rate of young men shows a pattern much like that for the homicide rate (see figure 6.1, middle panel). From World War II through the mid-1950s, when the relative number of young adults was declining, the suicide rate was virtually constant. Thereafter, as the relative number of young adults rose, the suicide rate increased until, by the late seventies, it was about three times that of the fifties. This pattern is consistent with my hypothesis: larger generation size contributes to greater psychological stress among young men.

The same pattern appears in the suicide rate for young females.[10] Although suicide is less common among young women than among young men, the post-World War II change in young women's suicide rates is very similar to that of young men—falling slightly to the mid-fifties and rising noticeably thereafter. This roughly parallel movement is symptomatic of stresses common to both men and women, stresses that, as we have seen, are reflected also in the swing in illegitimacy and marriage rates since World War II. As I have indicated, it can be attributed largely to the growing difficulty that young adults have encountered in realizing their life aspirations.

Younger versus Older Men: A Reprise. The generation size hypothesis is further supported by examining suicide and homicide rates for older adults. The change in the suicide rate among men aged forty-five to fifty-four contrasts noticeably with that of younger men (see figure 6.1, bottom panel). From World War II to the late 1950s, the rate for older men remained more or less constant. Since then, however, it has gradually drifted downward, while that for younger men has risen sharply.[11]

A comparison of homicide rates for older and younger men over this same period is not possible because for older men the victimization data are not good indicators of the commission of homicide. For the period since 1964, however, the murder arrest rates of younger and older men can be compared. Between 1964 and 1976 for men aged fifteen to twenty-four, the rate per 100,000 population more than doubled, from 7.2 to 14.7.[12] For men aged forty-five to fifty-four, the rate rose but much less than for younger men, from 3.5 to 4.7. These figures are consistent with those for suicide as well as for those for psychological stress in the preceding section. All three sets of data imply that during the last two decades younger men experienced a marked rise in stress as compared with older. This would be expected if differential movements in generation size had had an important effect on mental stress.

Fertility, Suicide, and Homicide. Demographer Martin O'Connell has pointed out that since the 1930s young men's suicide rates have a high negative correlation with women's fertility—in other words, when fertility has been high, suicide has been low, and vice versa.[13] Similarly, a study by British scholar William R. Lyster has noted a correlation between homicide and fertility.[14] Again, the relationship is inverse— higher fertility goes with lower homicide. The question arises as to whether these correlations are coincidental or involve real cause-effect relations.

The answer suggested by this analysis (and also by O'Connell regarding the suicide-fertility relationship [15]) is that fertility, homicide, and suicide all respond to a common causal factor—variations in the ability of young persons to achieve their life-style aspirations. As generation size declines and as the relative income of young adults improves, so too does their mental outlook. They are more likely to marry and to have children, and mental stress, as evidenced by crime and suicide, will decline. As generation size grows and as relative income deteriorates, the opposite will happen. Note that this is like the situation in chapter 4, where the correlation between fertility and variations in the uptrend of young women's work outside the home was attributed to the common impact of changing generation size.

Drinking and Drug Use.   Crime and suicide are relatively infrequent among the general population. But they are symptomatic of strains that are more widespread, as evidenced by the extent of drinking intoxicating beverages. As one would expect, the greater the stress, the greater the prevalence of drinking. Among persons in their twenties, the proportion of drinkers (i.e., nonabstainers) declined from 1947 to 1960 and turned up sharply thereafter.[16] In 1974, according to the National Institute of Alcohol Abuse and Alcoholism, 40 percent of male and 21 percent of female high school seniors had problems involving drinking, compared to 5 percent or fewer in 1960.[17] The rise in drug use in the sixties is in part probably also a response to increased mental stress among the young. Some of the increase in crime and suicide rates was undoubtedly due directly to the increase in drinking and in drug use, although mental stress was the root cause.

Accidents and the General Level of Young Adult Mortality
Most deaths among young adults are due to violence, that is, suicide, homicide, and accidents, with the latter—especially

motor vehicle accidents—the most frequent of the three. Increased mental stress is likely to raise the rate of mortality due to accidents for the same reasons that suicide and homicide rates go up.[18] The reckless driver who has a fatal car accident may be responding to pressures much like those that lead someone else to commit suicide. In addition, increased drinking and drug use is likely to raise the frequency of accidents. Between 1960 and 1973, the death rate among younger adults due to accidents rose noticeably. In 1974, following the imposition of the fifty-five-mile per hour speed limit, the death rate dropped, but in recent years, it has resumed its rise, almost reaching by 1978 its pre-1974 level.[19] According to J. Donald Millar, assistant director of the Federal Center for Disease Control, "what we're seeing . . . is an epidemic of deaths that has a direct relationship to drinking." [20]

This rise in violent deaths among young adults has produced a change in overall mortality for this group that contrasts noticeably with that for other age groups in the population. As a newly published report by the U.S. surgeon general points out, "the death rate for American teenagers and young adults has been rising since 1960 while death rates for every other age group have declined." [21] We have here dramatic testimony to the growth in mental stress among young adults.

Mortality and Generation Size. Relatively higher mortality may follow a large generation throughout its life. A recent study of younger adults notes that the mental stress associated with adverse economic and social conditions experienced by young adults may be transformed over time into physiological problems as well: "over the long term . . . chronic [social] stresses lead to chronic physiological disruption which contributes to major pathologies such as arteriosclerotic heart disease, hypertension, ulcers and asthma, and also to decreased immune resistance to infectious disease and probably cancer." [22] Thus, as a large cohort matures, it is likely to have a higher than expected incidence of diseases.

Political Alienation

When young adults find it easier to achieve their life-style aspirations, they are more likely to identify with the society in which they live; when they find it difficult, they are more likely to feel rebellious and alienated. Hence, one would expect that changes in generation size would be correlated with feelings of alienation. And this too turns out to be true. Intermittently since World War II, surveys have asked respondents whether they agree or disagree with such statements as:

"Sometimes politics and government seem so complicated that a person like me can't really understand what's going on."

"People like me don't have any say about what the government does."

An increase in the proportion agreeing with such statements is generally taken as an increase in feelings of political alienation. The change since the first survey (1952) shows a familiar pattern: until around 1960, alienation declined among young adults; thereafter, it rose (see figure 6.2).

Once again, however, we cannot presume that changing generation size fully explains this pattern. In the case of political alienation, Vietnam and Watergate undoubtedly played a part—possibly a dominant part.[23] But the adverse psychological impact of generation size possibly fostered a state of mind more responsive to these alienating events. A number of social scientists have suggested that the student activism in the 1960s was partly due to the impact of large generation size.[24] For example, Stanford political scientist Gabriel Almond has speculated that rebelliousness among teenagers is fostered by their relatively greater exposure to peers than to older adults since peers tend to reinforce extremist behavior.[25] Large generation size, of course, facilitates relatively greater exposure to peers.

FIGURE 6.2
Political Alienation among Younger Men

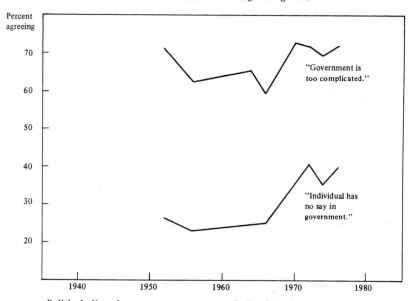

Political alienation among younger men declined in the 1950s and rose thereafter. This is shown by the proportion answering affirmatively to two statements about political alienation.

Source: Appendix Table 6.3, which also gives the precise statements.

Marked for Life? Some writers expect that, as the current large generation ages, it will become increasingly conservative. According to James W. Kuhn:

Having entered a tight labor market in the seventies, and all too often having had to accept work inferior to their training, such employees will savor the success that came late. From the perspective of each such late winner, the rewards are likely to appear the result of personal effort and merit. This perceived merit, taken with the lack of serious challenge by a younger generation and the increasing responsibilities bestowed by maturity may well turn the "Now" generation into one of the most self-satisfied, resistant-to-change sets of elders our nation has yet known.[26]

Although such a reaction is possible, in my view it is unlikely. The present large generation will remain relatively deprived throughout its career. It will continue to be subject to disproportionate psychological stress. And as noted, the scars of this stress are likely to be reflected in physiological condition as well. In my view, this generation, although forced by economic necessity into superficial conformity, will continue to be relatively unhappy, a prime target for those who feed on feelings of alienation.

Breakdown of Society?

In this chapter and the last, we have seen that the frequency of divorce, illegitimacy, crime, suicide, and alienation is higher among young adults who come from a large generation than among those who come from a small one. For a number of these conditions—notably divorce, illegitimacy, and crime—the frequency of such events relative to the population as a whole is also raised when a large generation reaches adulthood by what demographers call an "age composition" effect. This is

Social Disorganization

because the rates of divorce, illegitimacy, and crime are higher among young adults than other age groups. Hence an increase in young adults' share of the total population raises the average rate per thousand total population. For one or both of these reasons, therefore, rates of family and social disorganization have worsened in the period since 1960 as the post-World War II baby boom generation has reached young adulthood. This has created a growing feeling of social deterioration and concern for the future of American society.

The lesson of this and the previous chapter is that experience since 1960 is an unreliable guide to the long-term future. In the 1950s, the generation reaching adulthood was unusually small, and this contributed to a below-average frequency of such events; in the 1970s, the young adult generation was unusually large, and this contributed to above-average frequency. This shift from rates below to above average thus gives a distorted impression of the long-term outlook. Although the trend may be upward, it is unlikely to continue at rates as high as those recently experienced.

7

Stagflation

WE HAVE SEEN how changes in generation size have major repercussions on social health. As large generations reach adulthood, this leads to a growing frequency among young adults of such conditions as divorce, illegitimacy, crime, and suicide and creates a general feeling of deterioration in society.

Increases in generation size affect economic welfare as well. As large generations enter the labor market, they tend to raise the economywide rate of unemployment and precipitate more rapid inflation—to produce, in other words, what is popularly called "stagflation." This concurrent deterioration of economic and social conditions feeds on itself, confirming the impression that things are generally getting worse.

In this chapter, I make the connection between generation size and stagflation. This is, of course, not the whole story of stagflation—I am singling out the one aspect of the problem that is relevant here—but I think it is an important aspect.

The Problem

After World War II, students of economics were taught how the "Keynesian revolution" had provided economists with the proper prescription for the twin ills of unemployment and inflation. Appropriate federal policies could manipulate "aggregate demand"—that is, the combined spending of consumers, business, and government—to stabilize the economy with a low rate of unemployment and little or no inflation. The prescription for curing growing unemployment was to raise aggregate demand; for inflation, the opposite. And indeed, this prescription, which was translated into the Employment Act of 1946, proved to have substantial merit. As was noted in chapter 2, the unemployment rate of the last three decades has been markedly superior to any prior period of comparable length, and until the 1960s, inflation was kept within reasonable bounds.

To the discomfort of economists, however, a new phenomenon, "stagflation," has emerged in the last two decades. In theory, the problem had been one primarily of unemployment *or* inflation. However, in the 1970s, the average unemployment rate was higher than in the 1950s (6.2 percent versus 4.6 percent), as was inflation (prices rose at a rate of 6.8 percent per year compared with 2.1 percent in the fifties).[1] Thus, higher unemployment and more rapid inflation have gone together. The question is why, and where has theory gone wrong?

The Solution: An Overview

The answer that has found acceptance among many economists is that customary Keynesian economics placed exclusive emphasis on demand factors as the reason for and solution to the unemployment–inflation problem, whereas supply conditions that lie outside the reach of demand policies need also to be considered.[2] Supply conditions include such factors as unfavorable harvests, energy crises, and union–management wage settlements. In my view, and that of a number of other economists as well, the rise in the proportion of young to old in the working-age population has been another supply factor contributing to stagflation. Again, I do not claim that this is the sole cause of the problem, but it has been a contributing factor, probably an important one.

The gist of the argument is this.[3] When young persons are disproportionately plentiful in the labor force, as has been increasingly true since 1960, the average rate of unemployment tends to rise. Moreover, as the employment problems facing young men increase, young women increasingly move into the labor market, further aggravating the growth of unemployment. Conventionally, the government tries to counter increasing unemployment through monetary and fiscal policies that raise the demand for goods—for example, policies that make credit more accessible to prospective borrowers and tax-relief measures that leave households and businesses with more funds to spend. But these policies cannot cope with unemployment induced by supply conditions because the labor skills that are needed to expand output in response to the new demand are not the same as those available in the pool of unemployed workers. Therefore, the new demand for goods does relatively little to expand output and to reduce unemployment, but instead goes largely into increasing prices.

Using an individual firm as an example will clarify what goes on in the economy as a whole. Consider first how government policy works in the orthodox situation. The employment rate has been rising because private spending has dropped. To stimulate total spending, the government introduces tax-relief measures and relaxes constraints on credit. Consumers, finding themselves with more money to spend or with more readily available credit, increase their purchases at the retail

Stagflation

level. Department stores become depleted of stock sooner than expected and hasten to replenish it. The typical manufacturer, who, at the trough of the business recession, had been producing at, say, 75 percent capacity, finds himself in the pleasant position of unexpectedly receiving new orders. In response to these, he hires as many skilled older people to operate his machines and as many experienced supervisors as he can find. He also hires many unskilled, inexperienced younger workers to feed raw materials into the machines, wrap finished products, pack and stack them, and so on. If sales boom, he may build a new plant. This will mean, while the plant is being built, new demands in the construction industry for skilled and unskilled labor, and, when the plant is ready, new demands for manufacturing labor. In this way, government policies influencing the demand for goods translate into more jobs and lower unemployment rates.

Now consider what happens when unemployment is increasing because of a rise in the proportion of young to old in the working-age population, in other words, when the pool of unemployed workers is disproportionately made up of young unskilled workers. As a result of the same government policies, consumer demand rises, stocks are depleted, and stores place new orders. In response to the increasing demand, the manufacturer hires new workers, both skilled and unskilled, and expands production to capacity levels. As he does so, however, he finds it increasingly difficult to find experienced workers. But since plenty of young, unskilled, inexperienced workers are still swelling the unemployment rolls, the government continues to stimulate demand.

However, the manufacturer, having already hired all the experienced machine operators and supervisors he could find, cannot respond to that increased demand by building another factory—he cannot staff it with young, inexperienced workers. So how will he respond to the increased demand for his limited output? He will most likely raise his prices. At this stage, government demand policies translate into price increases rather than into unemployment reductions.

Demand, Supply, Unemployment, and Inflation

For some, the above illustration may be sufficient. For those who want more details and supporting evidence, let me develop the argument. In both parts of the example, unemployment is rising to unacceptable limits, and because of this, the government undertakes corrective action, stimulating total spending. But in one case, the government's action successfully reduces unemployment with little or no inflationary impact, while in the other, although the tide of rising unemployment is partly stemmed, the effect is largely to accelerate inflation.

The difference lies in the cause of the rising unemployment rate. In the first case, the cause is insufficient demand—a slump in spending that has caused firms to lay off workers. If the cause of increased unemployment is slackened demand, then stimulating demand will correct the problem.

In the second case, however, rising unemployment is caused by a rising proportion of younger to older persons in the working-age population, that is, a change in supply conditions. A prescription to stimulate total demand then complicates the problem by adding inflation to unemployment. To see this more fully, I will consider in turn:

1. How an increase in the proportion of younger to older persons in the working-age population raises:
   a. the average unemployment rate in the economy and
   b. the proportion of younger men and women in the pool of unemployed workers;
2. Why under these circumstances demand-stimulating policies will do little to reduce unemployment and will result instead in increased inflation; and

Stagflation

3. Some qualifications to the argument, including government policies that might ease, rather than aggravate, the problem.

The Effect of Age Structure on the Economy's Average Unemployment Rate. How does an increase in the proportion of young to old in the working-age population raise the economywide unemployment rate? There are several ways—an "age-composition effect," an "age-specific effect," and a reinforcing effect from the increased labor force participation of young women. Let us take each of these in turn.

To understand the age-composition effect, one must recognize that the economywide unemployment rate is an average of the rates for younger and older workers, with the rate for younger workers being higher than that for older because of their newness in the labor market, their job-seeking activity, the tentativeness of their job commitments, and so on (see chapter 2, figure 2.3). The economywide average, however, is not a simple average of the rates for younger and older workers; that is, to calculate the average, one does not merely add the rates for younger and older workers and divide by two. Rather, the rate is what is technically called a "weighted" average, in which the average is computed by weighting the rates for each group according to the group's share of the total labor force. The logic of a weighted average can be made clear by an extreme example: if the labor force were composed almost entirely of older workers, then the average unemployment rate for the labor force as a whole would necessarily approach that of older workers alone. If the labor force were composed almost wholly of younger workers, then the average unemployment rate would approach that of the younger group. Thus, the actual average at any given time lies between the unemployment rate for younger workers and older, and will be closer to one extreme or the other depending on the proportion between younger and older workers in the labor force. Because the unemployment rate for the younger group is higher than that for the older, if between two dates the share of the younger group rises, the average unemployment rate for the labor force as a whole is pushed up. This is called an age-composition effect because the change in the average results not from a change in the rate for either group, younger or older, but from a change in the shares of the two age groups in the total labor force. The following figures for male workers illustrate the approximate order of magnitude of the age-composition effect between the 1950s and the 1970s: *

	(1) (2) Males under 25		(3) (4) Males 25 and over		(5) Total labor force
	Unemployment rate	Proportion of labor force	Unemployment rate	Proportion of labor force	Unemployment rate
1950s	8.6	.13	3.1	.87	3.8
1970s	8.6	.22	3.1	.78	4.3

Note that the unemployment rate of the male labor force as a whole (column 5) rises between the 1950s and 1970s by 0.5 percentage points, even though the rate for each group, younger and older, stays the same (columns 1 and 3). The increase in the average occurs because the age distribution of the labor force (shown in columns 2 and 4) shifts toward the group with the higher rate. Thus, the rise in the proportion of young men increased the average male unemployment rate over the last two decades through an age-composition effect.

In addition to this age-composition effect, there was also an age-specific one. We saw in chapter 2 that a rise in the proportion of younger to older workers alters the supply–demand

* Because of the form of the age classification published for unemployment, the cutting line here has been taken at age 25 rather than 30. In the tabulation here and below, column 5 = [(1) × (2)] + [(3) × (4)].

Stagflation

balance in their respective labor markets, raising the unemployment rate of younger workers as compared with older. This adverse change in the rate for younger workers raises the economywide average unemployment rate even further, as shown below:

	(1) Males under 25	(2)	(3) Males 25 and over	(4)	(5) Total labor force
	Unemployment rate	Proportion of labor force	Unemployment rate	Proportion of labor force	Unemployment rate
1950s	8.6	.22	3.1	.78	4.3
1970s	11.4	.22	3.0	.78	4.8

Note how the average unemployment rate (column 5) is raised another 0.5 percentage points by the change in age-specific rates (columns 1 and 3), although this time the age composition of the labor force (columns 2 and 4) is unchanged. In reality, of course, this age-specific effect occurred concurrently with the age-composition effect. Hence, the male unemployment rate was raised in total by a full percentage point as the proportion of younger to older male workers rose between the 1950s and 1970s.

The illustration above is for males, but it applies to females too. The rise in the proportion of younger to older in the working-age population occurs for females as well as for males. Because unemployment rates for younger women are higher than those for older, the same type of age-specific and age-composition effects occur as for males, raising the average unemployment rate of females, though not, of course, in precisely the same amount.

In the case of women, however, an additional upward pressure is exerted on the average unemployment rate because the employment difficulties of young men have an impact on young women's work outside the home. In chapters 3 and 4, we saw that as the situation of young adults deteriorates the uptrend in labor force participation of young women is accelerated, partly because more of them remain single for a longer time and partly because those who are wives and mothers are under greater pressure to work outside the home. This accentuates the shift in age composition of the female labor force toward the young and thus puts further upward pressure on the economy's average unemployment rate. Younger females in fact have had an even greater labor force increase between the fifties and seventies than younger males and also have had a greater rise in their unemployment rate (see appendix table 7.1). This has added to the age-composition and age-specific effects, raising the economywide unemployment rate still further. A recent estimate of the combined effect of all these factors indicates that they have increased the economywide unemployment rate for men and women together by about 1.5 percentage points.[4]

Age Structure and the Pool of Unemployed Workers. The rise in the average unemployment rate is accompanied by another development of special significance—a change in the age-sex composition of the pool of unemployed workers. To show this, let us compare the age-sex mix of the unemployed at two dates, 1949 and 1978, when the overall rate of unemployment, about 6 percent, was fairly high by post-World War II standards, and government corrective action might be expected. As the illustration at the beginning of this chapter showed, the supply of experienced males—the group in the labor force that is, on the average, the most skilled—is crucial to the expansion of output because they cannot easily be replaced by younger men or women. Hence, the composition of the unemployed is given below in terms of how many younger men or women are unemployed for every ten older men out of work: [5]

Stagflation

	1949	1978
Males 16-24	5	10
Females 16 and over	6	20

Put simply, these figures show that in 1949 there were about five younger men for every ten older in the unemployment pool; by 1978, this proportion had doubled—there were ten younger men where there used to be five. For women, the growth relative to older men was even greater—from about six for every ten older men in 1949 to twenty in 1978. Thus, there were many more young men and women relative to older men in the unemployment pool of 1978 than there were in 1949.

The Mismatch Between Incremental Demand and the Unemployment Pool. We can now see more clearly why demand-stimulating policies cannot make a substantial dent in the unemployment problem when the primary source of unemployment is an adverse change in labor supply conditions. Demand policies trigger more spending, but the decisions about what goods will be purchased and produced are left to consumers and businesses. Clearly, the crucial question is how well the increased spending is tailored to the available labor supply mix. Government demand-stimulating policies raise the total level of spending, but these policies do little if anything to ensure that the composition of the increased demand will match that of skills in the unemployment pool.

The skill mix of the increased labor demand stimulated by government policies is most likely similar to the existing composition of *employed* workers. In other words, as demand rises, employers hire additional workers in about the same proportions of skilled to unskilled as they presently employ. Although these proportions are not rigidly fixed, it is reasonable to suppose that employers maintain a rough balance among different types of workers and further that that balance gives a reasonable indication of the proportions in which new workers will be hired as production expands. If this is so, then the extent to which unemployment can be reduced depends on the composition of workers in the unemployment pool compared with that of employed workers.

A comparison of the situation today with that in the past shows how the disparity between the composition of employed and unemployed workers has grown and, hence, how the effectiveness of government demand policy is limited. To simplify matters even further, let us lump together younger men and women as unskilled workers and treat older men as skilled (this is clearly an exaggeration, but it elucidates the main point). Consider first the proportions in 1949:

	Employed	Unemployed
Number per ten men aged 25 and over of:		
(a) Men 16-24	2	5
(b) Females 16 and over	5	6
(c) Total, (a) + (b)	7	11

In 1949, for every ten skilled workers employed, there were seven unskilled, and this would be the approximate proportion in which employers would hire as demand rose. Among those out of work, for every ten skilled, there were eleven unskilled. Although hiring at the rate of seven unskilled per ten skilled would not completely eliminate unemployment among younger men and women, it would go a long way. Of course, one would not expect unemployment to be wholly eliminated for any age-sex group because a certain amount of unemployment normally comes about through job-seeking activities (especially among the young), seasonal layoffs, and similar conditions.

What is of special interest here is how the gap between the composition of employed and unemployed workers has grown since 1949. Here are the 1978 figures:

Stagflation

	Employed	Unemployed
Number per ten men		
aged 25 and over of:		
(a) Men, 16-24	3	10
(b) Females, 16 and over	9	20
(c) Total, (a) + (b)	12	30

Today, for every ten skilled workers employed, there are somewhat more unskilled than in 1949—about twelve. This is especially true of women and reflects the long-term trend toward
a growth in "female" occupations relative to male. What is of
special importance here, however, is that today the skill mix of
the unemployed differs much more than it had in 1949. In the
unemployment pool, there are thirty unskilled workers for every
ten skilled, whereas hiring might be expected at the rate of
only twelve to ten. Relative to the proportions in which skilled
and unskilled would be expected to be hired, a much larger
surplus of unskilled workers was in the unemployment pool in
1978 than in 1949—or to turn it around, a much greater
scarcity of skilled workers existed. Thus, compared with 1949
present government policy to reduce unemployment by stimulating demand will much sooner encounter a shortage of older,
skilled workers, leaving unemployed large numbers of unskilled
workers. As a result, increased demand will improve the
average unemployment rate only slightly and will more quickly
increase prices.

The Noninflationary Rate of Unemployment. Implied in
the above is the assumption that labor supply conditions—the
proportion of young to old in the working-age population—limit
the effectiveness of government demand-stimulating policies in
reducing the unemployment rate without increasing inflation.
If the proportion of young persons is high, then aggregate demand policies will encounter this limit much sooner and will

precipitate more rapid inflation than when the proportion of young persons is low.

Let us call the minimum unemployment rate that government demand-stimulating policies can achieve without raising the rate of inflation, the "noninflationary" unemployment rate. If, for example, the noninflationary unemployment rate were 5 percent and the actual unemployment rate were 6 percent, then government policies to stimulate demand for goods could lower the actual rate to 5 percent without creating new inflationary pressures. But 5 percent is the limit. If the government further stimulates the economy, pushing the actual unemployment rate below 5 percent, prices will begin to rise more rapidly.

As we have seen, the noninflationary unemployment rate depends on the relative supply of younger workers in the economy—the higher their proportion, the higher the noninflationary unemployment rate. This rate is better thought of, not as a single value, but as a band around that value. However viewed, the noninflationary unemployment rate is not easy to estimate.

For our purposes, however, how much the increased proportion of younger workers has raised the noninflationary unemployment rate is more important. On this, we have the benefit of an estimate by economists Michael L. Wachter and Jeffrey Perloff of the University of Pennsylvania. According to their research, the noninflationary unemployment rate is about 1.5 percentage points higher today than it had been two decades ago.[6] In a society that is concerned when the actual unemployment rate increases by 1 percentage point, such as from 5 to 6 percent, this is a difference of considerable magnitude. The implication is that government policies to reduce unemployment by stimulating demand are likely to raise the rate of inflation considerably sooner today than they had in the 1950s.

How an Increase in the Noninflationary Unemployment Rate May Aggravate Inflation. An increase in the noninflationary unemployment rate does not necessarily have to lead to a higher inflation rate. If government demand policies were finely tuned to the changing limit set by new labor supply conditions, then excess spending could be matched by extra production. Unfortunately government policy has not been so finely tuned; moreover, it has sometimes had the effect of causing greater inflation along with higher unemployment.

To understand how this may happen, consider a "target rate" of unemployment, say, 5 percent, that is politically acceptable as "full employment." The government aims to maintain this target rate by demand-stimulating policies. Suppose further that a gradual growth in the proportion of young workers is pushing both the actual and the noninflationary unemployment rates from 5 toward 6 percent and that the government starts increasing the demand for goods in order to reduce unemployment to the target level of 5 percent. In this situation, the new spending caused by government policy can only push prices up. Hence, a higher rate of inflation and a higher unemployment rate will occur together.

Of course, it is extreme to assume that the government's stimulus to demand will do nothing to raise output and reduce unemployment. More realistically, the actual unemployment rate will end up at, say, 5.5 rather than at 6 percent; that is, it does not rise so much as the noninflationary unemployment rate. But the result is the same: concurrent growth in both the rate of unemployment and the rate of inflation.

Experience since the Fifties. Something of this sort appears to have happened in the last two decades. In the mid-fifties, judging from both low unemployment and low inflation, the noninflationary unemployment rate was probably about the same as the target rate. As we have seen, since then the increasing proportion of younger to older workers has increased the noninflationary unemployment rate by about 1.5 percentage points. However, through the late sixties, the government's target rate for demand policy remained essentially unchanged, and since then has apparently increased by something less than 1 percentage point.[7] Government policy appears therefore to have promoted spending greater than could be matched by new output and hence led to a higher inflation rate and a higher unemployment rate.

Since the 1950s, prices have moved consistently with this interpretation (see appendix table 7.2). In the fifties, the rate of price increase was only about 2 percent per year. Since then, it has swung upward as the proportion of younger persons in the labor force has risen, reaching an average rate of over 7 percent in the late seventies.

Why should the government response to an increase in the noninflationary unemployment rate be so slow—why has the target rate not been promptly revised upward as the noninflationary unemployment rate increased? The answer is, in part, because the problem has been inaccurately perceived. As I suggested at the beginning of the chapter, economic theory has only gradually abandoned the demand diagnosis of the unemployment–inflation problem and recognized the relevance of changing supply conditions. Partly, too, it is the difficulty of forecasting the noninflationary unemployment rate in such a way that takes account of the variety of changing supply conditions, not just the proportion of younger to older workers. It is easy to be a "Monday morning quarterback"—to say in retrospect that government policy should have been changed more rapidly. And finally, it is a matter simply of the government's moving slowly to raise the target rate because of the political repercussions of acknowledging that the minimum unemployment rate attainable through demand policy has gone up—to admit, for example, that traditional monetary-fiscal policies only can achieve an unemployment rate of, say, 6 percent.

Some Qualifications. It is worth repeating that this analysis is not intended as the final word on stagflation. Although altered labor supply conditions have contributed to the problem, and probably importantly, other factors have been at

Stagflation

work too. Energy crises; commodity market problems; and new health, safety, and environmental regulations have all played their part. Moreover, once rapid inflation has a foothold in the economy, it tends to be "built-in," or institutionalized, through such things as escalator clauses in wage contracts, and reinforced by its impact on expectations. If consumers and businesses think that prices are going to rise more rapidly in the future, then they are encouraged to "buy now." The increased spending engendered by the upward shift in price expectations further fuels inflation, confirming those expectations. It would be unrealistic, therefore, to suppose that further upward revisions of the target rate for government demand policies would have an immediate impact in reducing the rate of inflation, although it would probably help. Indeed, such revisions continue to take place. Government sources recently cited 6 percent unemployment as the minimum currently attainable through demand policies alone.[8]

I am not suggesting that raising the target rate of unemployment is the only response the government has to stagflation— that we must learn to live with more unemployment if we are to avoid greater inflation. It is possible for the government to use policies other than traditional aggregate demand policies.[9] On the labor supply side, the adverse effect on the skill mix of a rise in the proportion of younger to older workers might be partly countered by training programs for the young. Such supply-side policies might be accompanied by new government spending policies also aimed at the young, such as establishing a federal youth job corps that would employ young persons in public parks, forests, recreation areas, and so forth. These policies seek in effect to alter the composition of labor demand, as well as its level. Thus, the government can supplement the usual aggregate demand policies to reduce the actual unemployment rate below the noninflationary rate of unemployment without aggravating the inflation problem.

Stagflation and the Economic Security of Young Adults. By

aggravating stagflation, a large birth generation further undermines its economic outlook—unwittingly, to be sure. I have shown how a large generation's numbers when it enters the labor market impinge negatively on its earnings and unemployment experience relative to older workers (see chapter 2). We can now add to this the point that increased numbers coupled with inappropriate government policies also raise the rate of inflation. The feelings of economic insecurity engendered by inflation aggravate the lack of confidence already caused by employment and earnings difficulties among young adults. And their hopes for the future are further dimmed, increasing their reluctance to make such long-term commitments as marrying and having children. Moreover, the divisive pressures within marriage are intensified, and a psychology conducive to growing social alienation is fostered.

IV

Implications

8

The Future

THE GOOD NEWS is that things will get better; the bad news, it won't last. The swing in generation size that has had such a pervasive impact on social and economic conditions in the United States since 1940 is due to turn around again. The recent baby bust means that in the coming twenty years there will be a growing scarcity of younger men, eventually reaching the relative magnitude of the 1950s. The direct beneficiaries of this scarcity will be those born in the low birth rate era of the 1970s and, to a lesser extent, the 1960s. But America will benefit more generally as the scarcity of young adults brings with it an improvement in a variety of social and economic conditions, confounding the pessimists who have seen the dismal sixties and seventies as harbingers of worse to come.

After the turn of the century, however, the period of improvement may be followed by a return of two decades of deterioration; after that, another two decades of improvement. And so on into the future. The U.S. economy may be embarked on a self-generating cycle of around forty years—the swing we have witnessed from 1940 to 1980 appears to be the progenitor of more to come. Long fluctuations in the economy are not new phenomena—the record of population growth shows waves of about twenty years' duration from at least the mid-nineteenth century until World War II. But the swing since 1940 is about twice as long as those in the past and is due to a new source—a self-generating fluctuation in the birth rate. In this chapter, I shall take up first the story of the near-term future—the next twenty years—and then the long-term future.

The Good News

The Lucky Babies of the 1970s. Because the impact of changing generation size on the relative number of young adults has played such a dominant part in the experience of the last forty years, an attempt to predict the future must start there. Barring cataclysm, the prediction of the proportion of younger adults to older over the next two decades is simple and sure. It is simple because, as we saw in chapter 2, the relative number of young adults echoes the movement in the birth rate about twenty years earlier. The prediction is sure because virtually all those included in the projection are already born. Those who will be between fifteen and twenty-nine years of age in 1995 were born in the 1966–1980 period, for which, with the exception of the last year, we already know the birth record. Those who will be thirty and over in 1995 were born before 1966. Hence, the proportion that will exist between these two age groups in 1995 has already been shaped by the pattern of birth rates, and will be modified only in relatively small and predictable ways by mortality and immigration. This means that the outlook through 1995 for the relative number of young adults is known with unusual accuracy.

And what is the outlook? According to official estimates and projections, the proportion of younger to older men peaked around 1975 at about seventy-five younger per one hundred

FIGURE 8.1

The Prospective Effect of the Recent Baby Bust on the
Relative Number of Younger Men Compared with Older

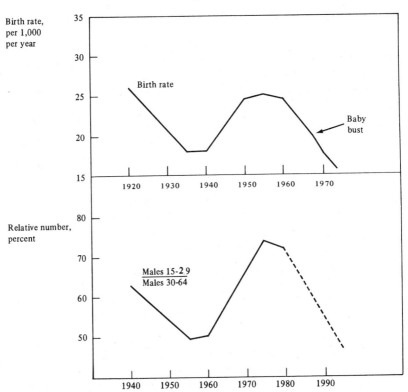

This figure, which extends figure 2.1 an additional fifteen years, shows that because of the baby bust of the last two decades (upper curve, right-hand segment) we can confidently project over the next two decades a substantial decline in the proportion of younger to older men in the working-age population (lower curve, broken line segment).

Source: Appendix Tables 1.1 and 2.1.

older, and has since declined slightly (see figure 8.1). From 1980 onward, a precipitous drop in the relative number of younger men will occur, as the baby bust of the sixties and seventies is echoed in sharply reduced numbers reaching adulthood. In the early 1990s, the proportion will reach about fifty younger men per one hundred older, down to the scarcity levels of the 1950s. Those who were born from the early 1960s onward increasingly will benefit from a growing scarcity of younger adults, but the full transition from surplus to scarcity takes time. By 1985, although the relative number of young adults will have decreased, the level will still be that of the late 1960s. The biggest beneficiaries of scarcity will be those born in the 1970s, who will reach adulthood in the 1990s, when the relative number of young adults troughs.

Social and Economic Effects. This new twist in age structure will cause a shift back toward the patterns of the 1940s and 1950s. Specifically, there should be:

1. An improvement in the labor market for young adults, with their earnings and employment prospects improving and their income increasing relative to older persons;
2. A gradual shift to earlier marriage, increased childbearing within marriage, and a consequent upturn in the birth rate;
3. A slowdown in the increase in younger women working outside the home, and an acceleration in work outside the home for older women;
4. A below-average growth in divorce and illegitimacy rates;
5. A decline in suicide rates of young men and women;
6. A lessening in crime and political alienation among the young; and
7. An abatement of stagflation.

In short, the next two decades should see a turnaround or improvement in a wide variety of social, economic, and political developments of the last decade or two. In some cases, because of other forces still at work, the underlying trend will continue, but the rate of change will be noticeably slower. In others, we may expect a reversal of recent patterns. None of these changes will take place overnight—for example, there will be no "rush to motherhood." But as economic pressures on the young slowly abate, a gradual but increasingly noticeable reversal of recent developments will occur.

The Past as Stepping-Stone to the Future. Why will a growing scarcity of young adults have such widespread repercussions? The answer follows from the record of the past forty years. As we move into the late 1980s and 1990s, those born in the recent low birth rate era will increasingly find, as they reach the labor market, that job openings are plentiful, wage rates relatively good, and advancement rapid. Their incomes will rise relative to older workers, and the unemployment rate among younger workers will fall. This, combined with the shrinking percentage of younger workers in the labor force, will lower the national unemployment rate and will ease inflationary pressures. The more favorable economic environment generally will reinforce the feelings of economic security that young adults are developing because of their favorable labor market experience. They will find it easier to satisfy their economic aspirations and to play their expected roles in life. Psychological stress will be reduced, and feelings of hopelessness or bitterness will be less prevalent.

How far will the reversal go? This is difficult to say, especially in view of the qualifications noted in the next chapter. As we have seen, however, by the mid-1990s the relative scarcity of young adults will reach a magnitude as great as or greater than that of the 1950s. The potential is there for a drastic reversal—for example, for a baby boom in the 1990s comparable to that of the 1950s. Should this occur, the lucky generation of the 1970s will in turn set the stage for a return to the relatively unhappy times of recent years. (But more on the longer term outlook in a moment.)

Although my view of the next twenty years is in the minority, I have some company on a few particulars. Economists Joseph M. Anderson, Michael L. Wachter, and Finis Welch separately argue that the growing scarcity of young adults in the next decade should result in a substantial improvement in their relative earnings and in a reduction in their unemployment rates.[1] Wachter also projects a reversal in the age pattern of growth in women's work outside the home.[2] Over the next decade, he foresees a rise in labor force participation rates for older women at a rate considerably higher than that for younger, in marked contrast to recent experience. Two demographers at the University of Michigan—economist Ronald D. Lee and sociologist David Goldberg—argue along lines similar to mine, that fertility may rise substantially.[3] One model of Lee's projects a rise in childbearing to a level not much less than the 1950s baby boom peak. And psychologist James Taylor predicts that a diminution in the numbers of young adults is likely to bring with it a reduction in "deviant" behavior, such as crime.[4] It is nevertheless true that I am probably alone in predicting such a sweeping reversal of recent developments.

Recently, a few straws in the wind suggest that some changes of the last decade or two may have been arrested. Since 1974, the rate of childbearing has remained about constant (see appendix table 3.2). A recent government publication reports that, from 1973 to 1977, "the growth in the divorce rate . . . slowed considerably especially in comparison with the period from 1967 to 1973."[5] Homicide rates among males aged fifteen to twenty-four, after peaking in 1974, dropped by more than 10 percent in the following three years (see appendix table 6.1). This early turnaround is of special interest because it is especially sensitive to the condition of teenagers, whose situation should improve earlier than that of persons in their twenties. It is too early to tell whether or not these developments are indicative of my prediction, but their timing is consistent with my argument.

The Future

Ups and Downs ad Infinitum?

Although the next twenty years look better, the longer term view is cloudier. The first two decades of the twenty-first century may see a return to the patterns of 1960–80. This, in turn, may be followed by another period of improvement. And so on, in forty-year cycles, into the future.

How the Cycle Works. This longer term projection is understood by recognizing that under the new conditions that followed World War II, a self-generating cycle in the birth rate may have been spawned. To see this, we can join together two earlier parts of the analysis:

1. The effect of the birth rate on the relative number of younger to older adults and
2. The effect of the relative number of younger to older adults on the birth rate.

As we have seen, a swing in the birth rate causes a corresponding swing about twenty years later in the proportion of persons aged fifteen to twenty-nine to those thirty to sixty-four (see chapter 2 and figure 8.1). Diagrammatically, the historical pattern can be represented as follows:

	1920-40	1940-60	1960-80	1980-2000
Persons $\frac{15\text{-}29}{30\text{-}64}$		Decline	Rise	Decline
Birth rate	Decline	Rise	Decline	

We also know that a decline in the relative number of young adults causes a concurrent rise in the birth rate because the relative income of young adults is improved, thus encouraging

137

marriage and childbearing. Conversely, an increase in their relative number causes a decline in childbearing (see chapter 3, especially figure 3.3).[6] This can be diagramed as follows:

	1940-60	1960-80
Persons $\frac{15-29}{30-64}$	Decline ↓	Rise ↓
Birth rate	Rise	Decline

When we fit these two diagrams together, a self-generating cycle is produced:

	1920-40	1940-60	1960-80	1980-2000	2000-2020
Persons $\frac{15-29}{30-64}$		↗Decline ↓	↗Rise ↓	↗Decline ↓	↗etc.
Birth rate	Decline↗	Rise↗	Decline↗	Rise↗	etc.

The last diagram shows that the young adult proportion in any given twenty-year period, on the one hand, is echoing the birth-rate movement in the prior twenty-year period and, on the other, is causing a concurrent birth-rate movement in the opposite direction. If one drops out the mediating role in the diagram of the young adult proportion, one sees directly that the birth rate in any given twenty-year period is causing an opposite movement in itself in the next period; that is:

	1920-40	1940-60	1960-80	1980-2000	2000-2020
Birth Rate ⟶	Decline ⟶	Rise ⟶	Decline ⟶	Rise ⟶	Decline ⟶ etc.

In other words, small generations tend to produce large generations, and vice versa. Thus, we arrive at a self-generating fertility movement that lasts forty years, if one counts both boom

The Future

and bust phases. This fertility movement is accompanied by a corresponding cycle in a wide variety of socioeconomic phenomena, reflecting variations in the relative income of young adults. This is the argument reduced to its simplest terms. The preceding chapters have fleshed out in cause–effect terms the bare bones skeleton shown here.

Even if fertility fluctuations and associated swings in socioeconomic conditions continue into the more distant future, they need not necessarily be so large as the cycle already experienced. But then again, they could be even larger. Can one say anything about the prospect of a dampening or magnification of future cycles? Frustratingly, little cyclical experience is presently available and is, in any event, probably better left to those more sophisticated in mathematical modeling.[7] It is worth noting, however, that when one enters in the diagram above the size of the birth-rate change in each of the three twenty-year periods experienced so far, no evidence of dampening is indicated:

	1920-40	1940-60	1960-80
Change in crude birth rate, percentage points[8]	−7.8	+6.7	−9.6

This is a slender reed, but for what it is worth, it suggests that future cycles may not be smaller than past.[9] However, in chapter 9, I shall note some factors that may dampen future cycles.

Because of my "good news" view of the next two decades, I am sometimes called an optimist. But the foregoing should make clear that I am an optimist only by comparison with those who see the next twenty years and beyond as an extension of the last twenty. For, although I see a turn for the better, I believe it is no more permanent than was the most recent turn for the worse. Rather, I see American society as possibly subject to waves of good times and bad. Politicians will doubtless seize on the "good times" as proof of their efficiency and will disclaim responsibility for the bad times. But in truth, the forces at work are larger than they.

What does a forty-year cycle in American life mean in terms of personal experience? If one thinks of adulthood today as about sixty years in length, then an individual would experience one and one-half cycles in his life. The small generations of the 1930s, for example, would witness as young adults good times from the end of World War II to 1960; in middle age, bad times from 1960 to 1980; and then in older age, the return to good times from 1980 to the start of the new century—not enough, perhaps, to confirm the existence of a cycle, but sufficient to instill the view that "the more things change, the more they are the same."

The Future Rate of Population Growth

The future size of the American population continues to be a subject of major concern. Some see the rate of childbearing stabilizing at or below replacement levels and American population consequently leveling off or declining in the course of time.[10] I suggest quite a different prospect: population growth in the future, as in the past, may be marked by sizable fluctuations, the length of which will be longer than in the past and the reasons different. The size of the American population would, therefore, continue to grow in roughly stepwise fashion.

Fluctuations: Past and Present. The past record of population growth contains little that warrants a projection of future stability. Historically, although the trend has been continuously upward, the rate of growth has varied considerably from one decade to the next. Before World War II, fluctuations in the rate of growth lasted around twenty years (see figure 8.2, top curve). Because Nobel prize-winning economist Simon Kuznets

FIGURE 8.2

Swings in Population Growth, Immigration, and Fertility

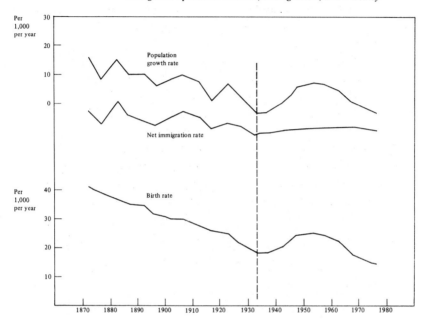

Long-period fluctuations in the rate of population growth (upper curve) have been occurring for at least a century, but the forty-year swing since the 1930s is about twice as long as those before the 1930s. The earlier fluctuations were due chiefly to immigration movements (middle curve); the recent one, to the baby boom and bust (lower curve).

Source: Appendix Table 8.1.

pioneered work in identifying and analyzing these movements, they are called "Kuznets cycles." [11]

On the surface, the record since World War II shows a continuation of Kuznets cycles—first an upsurge in the rate of growth (from the 1930s to the 1950s) and then a decline (from the 1950s to the present). Closer inspection, however, reveals noteworthy differences. The most recent swing is about twice as long as earlier ones, forty years compared with twenty. Moreover, the source of the swing is a movement in fertility— the baby boom and bust (see figure 8.2, bottom curve). In contrast, in the past the source of the swing was the immigration rate (see figure 8.2, middle curve).

In view of its different character, does the recent swing in population growth have any relation to past swings? The answer is yes. In both, an increase in the rate of population growth was initiated by a surge in labor demand triggered by an economic boom. As this boom progressed, the labor market grew progressively tighter. In the past, the resulting growth in wages and job opportunities ignited a wave of immigration from abroad, thereby raising the rate of population growth. This influx, in turn, helped to sustain the boom by generating new demands for housing and urban services generally.[12] The economic boom during and following World War II, however, was the first under conditions of restricted immigration. Moreover, it coincided with a situation in which, as we have seen, the declining birth rate from 1920 to 1940 had produced a growing relative scarcity of young workers in the post-World War II period. For the first time, therefore, a tight labor market redounded wholly to the benefit of young native American workers. This led, in turn, to earlier marriage, increased childbearing, and a rise in population growth that was due to a higher birth rate, not an increased migration rate.

The "New" Kuznets Cycle. Thus, the recent upswing in population growth, like past swings, was initiated by a major economic boom that raised the demand for labor. The altered

The Future

nature of the economy's labor supply, however, led to a much different type of population response—a rise in childbearing rather than in immigration. This, in turn, is the basis for a "new" Kuznets cycle.

The "old" Kuznets cycle was effectively put to rest by the Employment Act of 1946. With the federal government maintaining a high and growing demand for labor through monetary-fiscal policies, the main source of earlier Kuznets cycles—great fluctuations in the growth of the demand for goods and labor—was removed.[13] But the baby boom that occurred has led to a new cycle based on swings in labor supply conditions rather than in demand. Nothing was inevitable in this transition from old to new Kuznets cycles. If immigration had not been restricted in the post-World War II period, then the economic boom of that period would have led to a sharply increased flow of immigrants. The impact of labor market tightness on the incomes of young native American workers would have been much smaller, as would the stimulus to marriage and childbearing. With the maintenance in ensuing decades of a high and growing labor demand through government policy, immigration would have continued at high levels, and no substantial swing in the growth of labor supply would have occurred independently of labor demand conditions. In fact, however, the immigration legislation of the 1920s severely limited post-World War II immigration, the native population was the primary beneficiary of the postwar economic boom, and a major upswing in childbearing consequently occurred. Thus, a new cycle was set in motion, based, as we have seen, on a self-generating fluctuation in fertility of about forty years' duration, about twice as long as the historic Kuznets cycle. The twenty-year length of the expansion and contraction phases of the cycle reflects, of course, the time that must elapse before the impact of a turnaround in fertility is reflected in the economy's labor supply conditions, that is, in a reversal in the proportion of younger to older workers.

Note the contrast between old and new Kuznets cycles with regard to the relative roles of labor demand and supply conditions.[14] In past cycles, labor demand was the active initiating factor, and labor supply responded passively via immigration. Without swings in labor demand, no sizable swings would have occurred in immigration or in population growth. In the new Kuznets cycle, labor demand is passive, and swings in the composition of labor supply, echoing prior birth-rate swings, are the active factor. Thus, the relative roles of labor demand and supply have been reversed.

As far as future population growth is concerned, the new Kuznets cycle implies twenty-year spurts in population growth. In one period, the birth rate and population growth will surge upward; in the next, they will decline. Hence, the size of the population will grow in a more or less stepwise fashion. As I have mentioned, it is not clear whether or not future swings in the birth rate and population growth will be as great in the future as in the past forty years. Conceivably, the amplitude of the cycle might decrease very rapidly, in which case the new Kuznets cycle will join the old as a historical curiosity. But then again, it might be with us for some time to come. The prevailing view today is that American population growth is grinding to a halt. The lesson here is "caution!"

9

Conclusion

BEFORE the Industrial Revolution, the conditions of one's birth were perhaps paramount in determining one's lot in life. For the lucky few, born to high position, life was good. For the great mass of the population—for the "common man"—at best life was a meager, scraping thing.

The Industrial Revolution of the eighteenth century marked a sharp break with the past. The technology that came gradually into being over the course of the next two centuries—in Britain, the United States, northwestern Europe, and elsewhere—brought with it for the first time the possibility of a sustained advance in living levels for the population at large. Although pockets of misery remained, much of the American population in the 1970s lived as well as the wealthy of colonial times.[1] Not that the fortunes of birth no longer mattered—the high born still had more. But low or high birth was less important in a society where mass production meant mass consumption.

A new dimension was, however, added to the relation between personal welfare and the conditions of one's birth in pre–World War II industrial society. Modern technology brought with it a new source of instability in economic existence in the form of periodic depressions. Those whose year of birth brought them to the labor market in a period of mass unemployment often were scarred for life. The opposite was true of those who began their working life during economic booms.

In the United States, since the Employment Act of 1946, government management of the total demand for goods has sharply reduced the significance of this aspect of the "accident" of birth. True, economic recessions still occur, but today's recessions, unlike the massive depressions of the past, are hardly enough to ruin a start on a working life for large numbers of young people.

Yet the circumstances of one's birth still affect one's future personal welfare, although in a somewhat different way. Since World War II, under new conditions of government-sustained growth in labor demand and restricted immigration, a generation's fortunes have come to depend, as never before, on how numerous it is. If one is lucky enough to be born when the national birth rate is low—to come from a small generation—then one may look forward to a relatively bright future. If one has the misfortune of being a member of a large generation, then one's future is correspondingly dim. Moreover, variations in the fortunes of small or large generations have important consequences for the state of the economy and society generally. When small birth cohorts are young adults, social and economic conditions seem relatively prosperous; when large cohorts are young adults, life generally seems afflicted with malaise.

Generation Size and Personal Welfare: A Capsule View

The key links in the chain that connects a generation's size to its fortunes and to the state of the economy and society may be put as follows:

1. *Marriage, childbearing, and many other aspects of family formation and growth depend crucially on how the "typical" young couple assesses its "relative income," that is, the prospects for achieving the economic life-style to which they aspire.* The more favorable this assessment, the freer a cou-

Conclusion

ple feels to marry and raise a family, and the less favorable, the greater is the pressure on the young woman during the family-forming years to combine work outside the home with childbearing and childrearing. If young men and women can more readily realize their life-style aspirations and expected family roles, they will experience less psychological stress. But if they have difficulty in achieving their goals, such feelings as inadequacy, hopelessness, despair, resentment, and bitterness will be more widespread. These feelings will be reflected in a variety of social conditions—in an above-average frequency among young adults of illegitimacy, divorce, crime, alcoholism, motor vehicle accidents, suicide, political alienation, and so on. Such developments give a negative tone to the health of the society generally.

2. *A young couple's relative income depends in large part on the supply of younger workers relative to older when the partners are in the early working ages.* If younger workers are in relatively short supply, their earnings, unemployment experience, and rate of advancement will be favorably affected—all of which increases their relative income. If the opposite is true, the relative income of young couples will suffer.

3. *The supply of younger workers relative to older depends chiefly on their generation size, the national birth rate about twenty years earlier.* Small generations, hailing from low birth rate periods, such as the 1930s, are in short supply when working age is reached. Large generations, such as those born in the high birth rate era of the 1950s, are in excess supply.

The story of the past forty years is in part one of the impact of major swings in generation size on the economy and society. From World War II to around 1960, the small generations born chiefly in the low birth rate era of the 1930s were entering the labor market. Younger workers were in increasingly short supply, their ability to support their material aspirations improved sharply, their personal economic fortunes improved, and the general state of social and economic health was good.

But in the last decade or so, the labor market has been increasingly flooded with the offspring of the 1950s baby boom, the large generation that the previous small generation produced as a result of its good fortune. This large generation has had difficulty supporting its economic aspirations, its economic fortunes have suffered, and the general state of society and the economy has deteriorated. But just as the small generation of the 1930s reacted to its economic success by producing the large generation of the 1950s, so too the large generation of the 1950s, by virtue of its economic difficulties, has produced the baby bust of the last decade or so. As this small generation reaches adulthood, their fortunes will prosper like those of the young adults of the 1950s, and social and economic conditions generally should improve. But their good fortune may in turn sow the seeds of its own undoing by producing a new baby boom. And so it may go—small generations producing large, and large generations, small—with attendant swings in social and economic conditions. America seems embarked on a new type of fluctuation—a forty-year Kuznets cycle based on self-generating fertility swings. What many have seized on in the past two decades as proof of the breakdown of American society may turn out to be not a trend, but a fluctuation.

What Is Wrong with This Story?

In the preceding chapters, I have spelled out this argument in detail, presented supporting evidence, and dealt with a number of questions. Let me return briefly to a few of the questions, in terms that are likely to trouble most readers: even if what I say were correct as regards the experience of the past forty years, aren't things likely to be different in the future? My answer is, yes, to some extent. Hence, some modification of

Conclusion

the projection of chapter 8 is called for, but only in degree, not in kind. Let me take up some of the principal questions, particularly as they bear on the prospect for a new baby boom, because that is crucial to the notion that a new self-generating cycle may exist.

Sex Roles. The first question concerns the underlying assumption in my theory that traditional sex roles will persist. Some may feel that, while this assumption is true as regards the recent past, it does not apply to the future. New opportunities for women are opening up, and traditional notions of sex roles are being challenged. The "new woman" of the 1980s and 1990s may not want to have even two children, let alone three.

If the evidence showed that a substantial change would take place over the next two decades in attitudes toward sex roles, then the present projection of a baby boom should be revised downward. But as the evidence cited in chapter 1 indicates, most young adults today still have attitudes much like those of the past—they envisage working in typically male or female occupations, depending on their sex; they expect to have at least two children; and they anticipate that the wife will drop out of the labor force and devote herself to childrearing when the children are young. This does not deny the existence of dual-career families, where both husband and wife work continuously at full-time jobs. But most young adults today do not see that as their future. Although the proportion of dual-career families is likely to grow, there seems to be little evidence to justify projecting a dramatic increase in such families over the next two decades.

It is perhaps worth recalling that attempts to break down traditional sex-role attitudes have a longer history in countries like Sweden and the Soviet Union, but change there continues to be slow.[2] And there is good reason for this—the persistence of traditional sex-role indoctrination in the socialization of children. Trained from childhood to expect male and female adult roles essentially the same as have always existed, young persons could hardly be expected to embrace new life-styles in large numbers on reaching adulthood. This is not to say that none do, but as the data show, for the bulk of young adults, their expectations about their "proper" roles in life are much the same as ever.

Closely related to this is the criticism that my theory seems to imply "that the competitive economic advantage of reduced numbers is to be realized only by men. . . ." Might not an increase in economic opportunities among the young generally —women as well as men—"prompt young people to spend more on housing and other consumer items, and also attract women to better and to long-term career commitments?" [3] As far as it bears on the impact of scarcity on women's economic opportunities, this argument misses the mark. I have cited in chapter 2 a number of ways in which young women as well as young men benefited from labor market scarcity in the 1950s and suffered from labor market surplus in the 1970s. Indeed, as I have mentioned, I believe that both young men and young women will benefit from their growing scarcity over the next two decades and that this will encourage a mutually favorable assessment of the future and more willingness to marry and have children. It is also likely that a combination of affirmative action, changing sex-role attitudes, and growing scarcity of numbers will lead some women to career commitments in traditionally male occupations. But there is little, if any, evidence for believing that this will be a pervasive phenomenon in the next two decades. The position that seems most consonant with current evidence is that attitudes toward and performance of traditional sex roles will change only slowly—and then chiefly among highly educated persons—and that this will, at best, have a mild effect in dampening a new baby boom.

The Labor Market. A second question concerns the labor market for young adults. Are there, perhaps, new factors in the coming twenty years that may counterbalance the effects that scarcity produced in the 1950s?[4] Here, I think, one must concede a little more than on the preceding question.

First, illegal immigration, of only negligible importance in the 1950s, has increased. Although competition between "illegals" and native Americans is limited, a sustained inflow for two more decades equal to that of recent years would probably have some adverse effects on the labor market for young American men. And I believe such an inflow is likely to continue because the tightening labor market caused by a scarcity of young native American workers will stimulate it.[5]

Less certain is the effect of developments in the labor market for older workers. Because the prospective rise in the proportion of the aged population will have an adverse impact on the Social Security system, some have argued that older workers will be forced to stay in the labor force longer, a development aided by recent legislation against compulsory retirement. But the big increase in the percentage of the aged population will occur chiefly in the decades after the turn of the century, when those born in the post-World War II baby boom reach their sixties. Moreover, the average retirement age today is considerably below what it was in the 1950s. Hence, even if an increase in retirement age occurs in the next two decades, no greater competition is likely to result between younger and older workers than occurred during the last baby boom. In any event, the effect of a lengthening of retirement age on the opportunities of younger workers is problematical because older and younger men are mostly at opposite ends of the career ladder so that direct competition between them is limited.

More directly important to the opportunities of the young in the next twenty years is the prospective "downgrading" of the large generation that will be in the middle range of the working-age distribution by that time. Downgrading is a labor market phenomenon analogous to being "left back" in school. Whereas in the past a typical semiskilled worker could expect to advance to a skilled slot at, say, age thirty, the pressure of numbers will cause a rise in the average age of advancement, leading to a corresponding downgrading in the occupational distribution of those at any given age. This crowding would, in turn, act unfavorably on the opportunities of those further down the line—the smaller generation coming along.

Some downgrading will doubtless occur, but it will be modified by two circumstances. First, a certain proportion of the recent large generation has already been forced outside traditional career lines into marginal types of work. (For example, in the construction industry, some workers unable to obtain union membership, and thus entrée into the hierarchies of the building trades, have turned to self-employed nonunion work.) Second, firms seeing large numbers coming down the line, will be pressured to make room at the top, and some accommodation seems likely. Nevertheless, downgrading will probably have some negative impact on opportunities for younger workers, perhaps more so in the eighties than in the nineties, by which time the gaps in both age and numbers between the large and small generations will have widened.

These negative changes in the labor market for young men may be offset by a possible increase in the educational advantage of the small generation workers over their large generation predecessors. In the United States, the long-term trend toward increased schooling has meant that younger workers are generally better educated than older, although the size of the differential has varied. During the 1970s, for example, there was a marked interruption in the historical trend of rising male college enrollment rates. As a result, those aged twenty-five to twenty-nine today have no educational advantage over those aged thirty-five to forty-four. This interruption may have been caused by two factors, both related partly to large generation size. First, the adverse earnings impact of large numbers has been greater for those with a college education than for those without. The financial payoff from a college education has been declining, therefore reducing the incentive to obtain a college degree.[6] Second, parents of college-age children have recently been experiencing a financial squeeze, making it harder for them to pay for a college education. This is because an increasingly large proportion of households has found itself with two or more children simultaneously needing support for a college education as the baby boom generation reached college age.[7] Both these developments will be reversed in the next twenty years, as the baby bust generation reaches college age. A resumed increase in college enrollment rates among young men should result, along with a consequent growth in their educational advantage over their large generation predecessors. This should improve the competitive position of the young and help to offset some of the unfavorable developments previously mentioned. However, the later age of labor market entry for young men implied by this development would tend to defer somewhat the timing of a post-1980 upswing in marriage and childbearing.

Affirmative action programs are likely to be increasingly important in the future, but their net effect on the job prospects of young adults is unclear. The principal beneficiaries of such programs are likely to be younger women and younger minority group members, men and women. On the other hand, white males, younger and older, who will be under pressure to make room for those favored by these programs are likely to be adversely affected. The net balance of these effects on young persons is uncertain.

My projection is that a larger proportion of young women, wives included, will be working outside the home in the next two decades than in the 1950s but that the rate of increase will slow down. Is such a projection compatible with a new baby boom? In answering this question, it is helpful to think of the rate of childbearing among young wives as an average of two rates—the rate for wives who are not in the labor force and that for wives who are. For those who are not, nothing prevents the rate of childbearing from rising to levels like those of the 1950s. For those who are, new developments, such as the availability of child-care centers and maternity leaves, make an increase in childbearing rates more compatible with work outside the home than in the past. So too would an increase in the proportion of working women holding part-time rather than full-time jobs. Such a shift is likely because an improved job situation for young husbands would reduce the pressure on young women to supplement the household's income. Even for dual-career households, the possibility of higher fertility is not precluded because if a large family is desired the extra family income brought in by two full-time workers can be used to purchase full-time help for child care and housework. Thus, although the increase in young wives' labor force participation may have a dampening effect on a baby boom, it is compatible with considerably higher childbearing than presently exists.

Currently, many young adults favor two children as their ideal family size, and some readers may wonder why young adults should feel differently a decade hence and have more than two, even if they can afford it. I believe that a relaxation of economic pressures is likely in several ways to encourage a number of young couples to have more than two children. First, if they are feeling less pressure to limit family size, they may, in Norman B. Ryder's words, be "less vigilant" in their contraceptive practice, and unintended births may occur.[8] Second, if economic circumstances are favorable, preferences in terms of the boy-girl makeup of the family may push couples beyond two children. In a typical distribution of one hundred two-child families, half of the parents have only boys or only girls. Because parents have a strong preference for a child of each sex, some couples with children of only one sex are likely to have one more child, and they are more likely to do so if their economic circumstances are comfortable. The probability is that only half of those couples that try will succeed;

some of those who fail may decide to give it yet one more try, if economic circumstances permit. As somewhat larger families become prevalent, family-size desires are likely to increase—in the 1960's baby bust a change in actual childbearing behavior similarly foreshadowed a change in family-size preferences.[9] In ways such as these, improved relative income leads young adults to increase family size.

Confidence in the Future. The recurrent energy crises of the last decade have had a negative impact both on household budgets and on public confidence. Have they altered conditions so much that young couples of the eighties and nineties will hesitate to have children, even if their economic circumstances are relatively affluent? The answer, I think, is that the American economy is already starting to adjust to a new energy environment, one similar to that of Europe in recent decades. Higher energy prices have been partly offset by a switch to fuel-saving techniques by both industry and consumers—witness, for example, the shift to smaller and higher fuel economy automobiles. This and similar adaptations, such as substitute domestic energy sources, will continue. It seems reasonable to suppose that by the late eighties and nineties energy crises will no longer be a recurrent headache.

Similarly, there are those who argue that an inflationary environment has an unsettling effect on long-term decisions of couples and that a continuation of recent rates will seriously counter any forces favoring high childbearing. Closely related to this is the supposed negative impact on family formation of housing prices, which in recent years have increased even more rapidly than prices generally.[10] This disproportionate rise in housing prices is itself partly the effect of inflation, as young adults have increasingly seized on homeownership as a hedge against inflation.[11] It is also partly due to the high demand generated by the large generation currently reaching the age of homeownership. But as I have indicated, the shift toward a smaller proportion of young adults in the population over the next two decades should moderate inflationary pressures. In particular, the pressure on housing prices should be relieved, both directly by reducing the pressure of numbers on the housing market and indirectly by easing inflationary pressures generally. Nevertheless, one cannot easily dismiss the prospect that substantial inflation may continue and add a negative element to the childbearing picture that was absent in the fifties.

It is worth recalling, however, that factors that tend to offset the effect of generation size are not peculiar to the next two decades. Thus, it is not difficult to think of circumstances that might have dampened the last baby boom. After World War II, for example, the atomic bombings of Hiroshima and Nagasaki were fresh in people's minds. Bruce Bliven, writing in 1950, observed that "half a century ago, mankind, and especially the American section of mankind, was firmly entrenched in the theory that this is the best of all possible worlds and getting better by the minute. . . . Today, we have lost this faith and are frightened to death—of war, atom bombs, and the looming prospect of a general brutalization and deterioration of the human species." [12] With such feelings, young adults in the 1950s might have been expected to question the wisdom of bringing offspring into the world. Yet whatever its effect, the largest upsurge in fertility that this country has ever experienced was not prevented.

Forecast Revised. Taking everything together—sex roles, labor market developments, and other factors—it seems likely that the upswing in fertility over the next two decades will be somewhat more gradual and smaller in amplitude than one might otherwise project on the basis of the effect of generation size alone. Nevertheless, I think that it will show up clearly as a pronounced movement. By the mid-1990s, the prospective scarcity of young adults will be as great as during the 1950s baby boom, and although we probably will not replicate the earlier baby boom to the full extent, we should be well above replacement fertility levels.

Conclusion

In dealing with the future, I have put in the foreground the question of a prospective baby boom because it is critical to the argument that a self-generating process may now exist that will lead to recurrent cycles in the longer term future. If negative factors of the sort just discussed prove so powerful that a new baby boom does not materialize, the new Kuznets cycle would come to an end. As with the old, so with the new—with Moses Abramovitz we might say, "The Kuznets cycle in America lived, it flourished, it had its day, but its day is past. . . . *Requiescat in pace.*" [13]

Maybe There's Only Good News. But that is the longer term future. As far as the next two decades are concerned, the validity of my forecast does not hinge on a new baby boom. Even if low and declining rates of childbearing persist and young wives continue to work outside the home in increasing proportions, the economic pressures on young adults will be substantially reduced. As their numbers decline compared with older workers, their relative earnings and unemployment situation will improve, good jobs at decent wages will be easier to find, and movement up the career ladder will be more rapid. The diminution in the share of young workers in the labor force will also help to moderate the problem of stagflation. The reduction in economic pressures generally should decrease the prevalence of psychological stress and mental depression among young adults. Marital strains should be lessened, correspondingly reducing the steep rise in divorce rates. Suicide rates should turn down, and crime rates improve as the pressures felt by young adults lessen. Thus, the outlook for the next two decades is for better times, whether or not a baby boom occurs. And should it fail to materialize, the children of that generation would not experience a turn for the worse as they reached working age. Without the baby boom, there is only good news.

Trends and Cycles. Even if a new baby boom does occur, we should not suppose that the next cycle will be a carbon copy of the past. I have already indicated that the next baby boom is likely to be milder than that of 1940–60. If this proves true, its adverse impact on social and economic conditions in the two decades after 2000 would be correspondingly lessened.

Even if the new baby boom replicates that of the fifties, there would still be differences in future cycles. When I argue that some observers have confused fluctuations with trends, I do not mean to suggest that no real trends are at work. The social and economic environment twenty years hence will differ from that of today. There will be more material affluence as living levels continue to rise. Levels of education will be higher, and entry into the labor force will come later. The proportion of females (younger and older together) working outside the home will continue to rise, as will the proportion of female to male workers. Even if the rise in the divorce rate slows down in the next two decades, the level twenty years hence would still be higher than at present. At a minimum, such trends mean that there will be a difference in the base from which the next cycle will be launched. But they also mean that the nature of the cycle itself will be altered. Both the timing and the amplitude of the next baby boom are likely to be affected by such trends as later labor force entry on the part of young men and increased labor force participation among young women. Even if a self-generating cycle does exist, one must not suppose that future cycles will simply repeat the past.

Other Complaints. The present analysis is sometimes criticized on the ground that it projects future cycles on the basis of only one (supposed) cycle actually experienced, that since 1940. But clearly, one cannot go back further because, as shown in chapter 2, the conditions that gave rise to the present cycle did not exist before. Moreover, one cycle is a better basis for prediction than half a cycle—and that, in my opinion, is what is used by those who are extrapolating conditions since 1960 into the future.

Conclusion

Even though I have departed from traditional economic theory by bringing the formation of aspirations into my analysis, some still reject my theory as too "economic." And in truth, even when discussing aspirations I have stressed the importance of changing economic conditions as a determining factor. My defense here is simply that my concern is to explain changes in behavior over time, and for this purpose, economic factors are especially pertinent because they usually change more rapidly than other causal factors. For example, at a given point in time, religion undoubtedly plays a part in explaining differences among couples in size of family; and physical attraction, in explaining marriage behavior. Over time, however, the distribution of the population by religion and physical attractiveness changes only slowly, if at all, while economic circumstances change substantially. Hence, if one is to explain historical behavior, the emphasis falls logically on economic factors.

An objection likely to be raised by some of my fellow scholars is that, judged by contemporary social science standards, the methodology used here is too simple. Undoubtedly, more intensive statistical analysis on each of the topics taken up here would be helpful. But a large number of subjects are covered, and the historical data for them differ substantially in their frequency and reliability. Moreover, communicating with different academic disciplines and with the public would clearly be impeded by extensive use of formal statistical techniques. In any case, I am not claiming that the treatment of any of the topics discussed here is complete—generation size is certainly not the only factor at work, as I have indicated in the treatment of individual topics. Indeed, to treat fully every topic, a separate book on each would have been required.

The plausibility of the present argument, however, does not rest simply on whether the analysis of each subject is exhaustive or the methodology the most advanced. The whole *is* more than the sum of its parts. What is striking, I think, is the way in which a variety of developments over the past four

159

decades—developments often regarded as puzzling or surpris-
ing—form a coherent picture when approached from the present theoretical viewpoint. This viewpoint, moreover, is consistent with theoretical notions in several different disciplines,
theories that have heretofore led largely independent lives—
in economics, the work of such scholars as Duesenberry and
Modigliani on "relative income"; in sociology, Durkheim,
Merton, and others on anomie and related topics; in social
psychology, Stouffer and Festinger on relative deprivation and
cognitive dissonance.[14] The consistency of the present theory
with a wide range of behavior and its consonance with a
variety of theoretical perspectives are what give special cre-
dence to it.

Other Times, Other Places, Other Topics. But certainly
more testing is needed, and there is plenty of opportunity.
Along with more rigorous statistical techniques, the best way
to test the present theory further is to explore its applicability
more widely.

As far as earlier times are concerned, the door is unfortunately closed. The effect of generation size has come to
predominate in the United States only since 1940, when other
labor market conditions that had historically been quite variable were largely stabilized.

But we have the option of studying other places. The less
developed countries are ruled out because childbearing levels
are high and only partly regulated, and institutional conditions
are quite different. Before 1940, however, many developed
countries experienced a decline to low levels of fertility. Some
preliminary studies indicate that since 1940 childbearing may
have been influenced by relative income or generation size in
Canada, Australia, the United Kingdom, Denmark, Norway,
Sweden, Finland, Belgium, France, and the Federal Republic
of Germany.[15] Indeed, French demographer Jean Bourgeois-
Pichat has suggested that a self-generating fertility fluctuation
may be a common feature of developed countries once they

Conclusion

reach a low fertility level.[16] Nothing has been done, however, to explore the possibility that a broad set of interrelated social and economic changes such as occurred in the United States took place in these countries. Actually, the effect of generation size most likely will take different forms in different environments because of national differences in labor markets and other institutions. Both the Soviet Union and Japan, for example, are experiencing a substantial scarcity of young workers, and the effects of and adaptations to such a scarcity in these societies presumably would be quite different from those in the United States. Even within Western Europe and its overseas descendants considerable variability exists in institutions —the United States is at the free market end of the spectrum of economic systems, and the specific form of the present analysis clearly presupposes a free market economic system.[17] Hence, one would not expect a mere replication of American experience even by these developed countries.

In seeking new applications, one does not have to go beyond American borders. Minority groups, such as blacks and those of Spanish-speaking origin, provide another testing ground. My analysis has concentrated on the population as a whole, and the question exists about whether or not it applies to various subgroups. I believe that it does—for example, the baby boom and baby bust were widely diffused throughout the population.[18] Among minority groups the proportion deliberately regulating fertility is usually lower, and the level of their fertility is consequently higher. But the same economic pressures are at work as in the general population.

New applications also may involve other subjects. An obvious possibility is education. I noted earlier that variations in male college enrollment rates may reflect generation size. It seems likely that the same applies to female enrollments. If marriage prospects are good, this should retard the rise in female enrollment rates; if marriage prospects are poor, female enrollment should be accelerated. A hasty glance at experience supports this. In the 1940s and 1950s, the public was concerned about teenage women dropping out of school to get married. In contrast, in recent years female school enrollment rates have been rising rapidly. This rise may slow down in the next two decades if marriage options take a turn for the better.[19]

Many organizations, such as corporations and government agencies, operate with a set of internal rules, some of which are age related. Job hiring specifications, for example, may include an age stipulation, and promotion reviews may be geared to tenure. When similar rules are fairly widespread, large changes in the age distribution of the population create problems in the application of these rules, and organizations are pressured to make rule changes.[20] Again, some evidence indicates that such adaptations have occurred. For example, in the post-World War II era, the scarcity of young unmarried women forced the New York headquarters of International Business Machines to abandon its long-standing policy that a young secretary lost her job when she married. Airlines made a similar change in their requirements for the employment of stewardesses.[21] As the impact of the baby boom was felt on college enrollments, traditional tenure provisions for assistant and associate professors were waived, and advancement to full professorship was accelerated. Studying such institutional responses to demographic pressures might open up a whole new set of generation size effects.

What Can Be Done?

Suppose my diagnosis is correct, that American life is now subject to a sizable self-generating cycle in social and economic conditions of around forty years' duration. Must we then

Conclusion

simply live with it? Or can we do something to reduce substantially or perhaps to eliminate this new Kuznets cycle?

From what I have said, I obviously do not think that we are fated to a mechanical repetition of cycles. I have pointed out, for example, that various trends will alter the shape of the next cycle. But what about public policy? Do I have a suggestion to offer?

I must honestly answer no. In economics, as in medicine, a division of labor exists between those who seek the causes of disease and those who, once the causes are known, search for cures, and I make no claim to the necessary expertise on public policy. The general lines of such policy are, of course, fairly clear. Since the basic problem concerns the disproportion between younger and older workers, the solution, in general, is to devise policies that narrow the differences between the two groups, particularly as regards skill levels. Such policies might encompass formal education, special training programs, and incentives for firms to vary their hiring and promotion practices. Undoubtedly, we can profit from a study of the institutions and policies of other developed countries, where the impact of generation size has had different effects.

Of course, the swing in the supply of younger workers relative to older reflects an underlying swing in the birth rate, and this raises the question of whether or not we should direct public policy to regulating childbearing. My feeling on this is no, as it would be a major departure from our leaving choice of family size to the individual; in any event, we know little about how to influence childbearing through public policy. In those developed countries that have tried, the policies seem to have been notably ineffective.

Let me conclude on a note of hope. Today, as in the past, the economic fortunes of different generations are shaped largely by circumstances beyond their control. Before World War II, one's luck was determined by whether one came of age in an *economic* boom or bust; since then, it has largely depended on whether one comes from a large or small generation—from a *baby* boom or bust. We have learned how to moderate the shifts in fortune due to great swings in labor demand. It seems reasonable to suppose that, in time, we will learn how to control variations in the fortunes of generations caused by their relative size.

Appendix Tables

APPENDIX TABLE 1.1

Crude Birth Rate, 1915-78

Period	(1) Rate
1915-20	26.1
1920-25	25.0
1925-30	21.5
1930-35	18.3
1935-40	18.3
1940-45	21.2
1945-50	24.5
1950-55	25.2/24.8
1955-60	24.6
1960-65	22.2
1965-70	18.1
1970-75	15.9
1975-78	15.0

Sources: 1915-55: Simon Kuznets, "Long Swings in the Growth of Population and in Related Economic Variables," *Proceedings of the American Philosophical Society* 102 (1958), p. 37, table 1, column 4, table 3, column 5; p. 41, table 5, column 7; and p. 43, table 6, column 5 (underlying unrounded quinquennial estimates were used). 1950-75: U.S. Bureau of the Census, *Current Population Reports*, series P-25 (Washington, D.C.: U.S. Government Printing Office, September 1977), no. 706, p. 7. 1976-78: National Center for Health Statistics, *Monthly Vital Statistics Report* 27 (8) (Hyattsville, Maryland, 15 November 1978), p. 1. The break in the series in 1950-55 is due to the shift in sources.

APPENDIX TABLE 2.1

Male Population Aged 15-29 as Percentage of That Aged 30-64, Actual 1940-75 and Projected 1980-95

	(1)	(2)	(3)
	Male Population Aged		**Ratio, 15-29**
Date	**15-29** **30-64** (thousands)		**30-64** (percent)
1940	17,442	27,664	63.0
1950	17,216	31,761	54.4
1955	16,772	33,781	49.6
1960	17,794	35,478	50.2
1965	21,151	36,295	58.3
1970	25,262	38,115	66.3
1975	28,793	38,908	74.0
		Projected	
1980	30,426	42,184	72.1
1985	29,717	46,210	64.3
1990	27,626	50,585	54.6
1995	25,864	54,759	47.2

Sources: U.S. Bureau of the Census, *Current Population Reports*, series P-25 (Washington, D.C.: U.S. Government Printing Office), as follows—for 1940-55—no. 98, p. 115; for 1955—no. 265, p. 25; for 1960—no. 286, p. 42, series C; for 1965—no. 519, p. 20; for 1970-75—no. 614, pp. 11-16; for 1980-95—no. 704, pp. 40-60, series II.

APPENDIX TABLE 2.2

Median Income of Year-Round
Full-Time Income Recipients in
Specified Sex-Age Group as
Percentage of Average, 1955
and 1977

Age group	(1) 1955	(2) 1977
All males	100.0	100.0
14-19	43.3	40.1
20-24	77.7	65.0
25-34	101.7	93.8
35-44	106.5	111.9
45-54	106.0	113.0
55-64	93.4	104.0
All females	100.0	100.0
14-19	83.8	63.9
20-24	101.2	85.1
25-34	104.4	108.3
35-44	104.5	105.3
45-54	102.4	103.7
55-64	93.6	100.4

Sources: 1955–Bureau of the Census, *Current Population Reports*, series P-60, no. 23 (Washington, D.C.: U.S. Government Printing Office, November 1956), table 3, p. 13. 1977–Bureau of the Census, *Current Population Reports*, series P-60, No. 116 (Washington, D.C.: U.S. Government Printing Office, July 1978), table 10, p. 16.

APPENDIX TABLE 2.3

Percentage Employed of
Labor Force in Specified Sex-
Age Group, 1950 and 1978

Age Group	(1) 1950	(2) 1978
All males	94.9	94.8
16-17	86.7	80.8
18-19	87.7	86.8
20-24	91.9	90.9
25-34	95.6	95.7
35-44	96.4	97.2
45-54	96.0	97.2
55-64	95.1	97.3
All females	94.3	92.8
16-17	85.8	80.5
18-19	90.2	84.7
20-24	93.1	89.9
25-34	94.3	93.3
35-44	95.6	95.0
45-54	95.5	96.0
55-64	95.5	96.8

Sources: 1950–*Manpower Report of the President* (Washington, D.C.: U.S. Government Printing Office, April 1974), p. 272. 1978–*Employment and Earnings*, Bureau of Labor Statistics, 26 (1) (Washington, D.C.: U.S. Government Printing Office, January 1979), pp. 156-57.

APPENDIX TABLE 3.1

Annual Average Total Money Income of
Families with Head 14-24 Years Old as Relative of
That of Families with Head 45-54, 1957-78
($ of 1964 purchasing power)

(1)	(2)	(3)	(4)	(5)	(6)
Head 14-24 years old		Head 45-54 years old		Relative income (2) ÷ (4)	
Period	Income	Period	Income	Year	Percent
1950-54	$ 3,596	1947-53	$ 4,917	1957	73.1
1951-55	3,687	1948-54	5,028	1958	73.3
1952-56	3,814	1949-55	5,203	1959	73.3
1953-57	3,956	1950-56	5,439	1960	72.7
1954-58	4,033	1951-57	5,641	1961	71.5
1955-59	4,132	1952-58	5,830	1962	70.9
1956-60	4,210	1953-59	6,011	1963	69.7
1957-61	4,222	1954-60	6,204	1964	68.0
1958-62	4,241	1955-61	6,404	1965	66.2
1959-63	4,282	1956-62	6,602	1966	64.8
1960-64	4,414	1957-63	6,795	1967	65.0
1961-65	4,625	1958-64	7,019	1968	65.9
1962-66	4,824	1959-65	7,288	1969	66.2
1963-67	5,032	1960-66	7,564	1970	66.5
1964-68	5,289	1961-67	7,874	1971	67.2
1965-69	5,485	1962-68	8,205	1972	66.5
1966-70	5,520	1963-69	8,566	1973	64.4
1967-71	5,506	1964-70	8,872	1974	62.1
1968-72	5,523	1965-71	9,175	1975	60.2
1969-73	5,532	1966-72	9,507	1976	58.2
1970-74	5,492	1967-73	9,814	1977	56.0
1971-75	5,377	1968-74	10,017	1978	53.7

Source: Richard A. Easterlin, "Relative Economic Status and the American Fertility Swing," in Eleanor Bernert Sheldon, ed., *Family Economic Behavior: Problems and Prospects* (Philadelphia: J. B. Lippincott Company, 1973), p. 185, table 12, cols. 5 and 6. Recent data were kindly provided by Dr. Campbell Gibson, Bureau of the Census. The choice of dates used in constructing the relative income measure is explained in ibid., pp. 182-86.

APPENDIX TABLE 3.2

Total Fertility Rate, 1940-77; Relative Employment Experience of Young Adult Males, 1940-55; and Ratio of Males Aged 30-64 to Males Aged 15-29, 1940-75

Year	(1) Total Fertility Rate	(2) Relative Employment Experience	(3) Ratio, Males $\frac{30\text{-}64}{15\text{-}29}$
1940	2.30	−10.2	1.586
1941	2.40	−8.6	
1942	2.63	−5.8	
1943	2.72	−4.4	
1944	2.57	−3.2	
1945	2.49	−2.9	
1946	2.94	−3.3	
1947	3.27	0.4	
1948	3.11	3.5	
1949	3.11	6.1	
1950	3.09	7.2	1.840
1951	3.27	7.0	
1952	3.36	7.2	
1953	3.42	7.5	
1954	3.54	7.6	
1955	3.58	6.7	2.014
1956	3.69		
1957	3.77		
1958	3.70		
1959	3.71		
1960	3.65		1.994
1961	3.63		
1962	3.47		
1963	3.33		
1964	3.21		
1965	2.93		1.716
1966	2.74		
1967	2.57		
1968	2.48		
1969	2.45		
1970	2.47		1.509
1971	2.28		
1972	2.03		
1973	1.90		
1974	1.86		
1975	1.80		1.351
1976	1.76		
1977	1.83		

Sources: Total fertility rate: Richard A. Easterlin, *Population, Labor Force, and Long Swings in Economic Growth* (New York: National Bureau of Economic Research, 1968), p. 247, col. 4, and Campbell Gibson and Martin O'Connell, U.S. Bureau of the Census. Relative employment experience: Richard A. Easterlin, "Relative Economic Status and the American Fertility Swing," in Eleanor Sheldon, ed., *Family Economic Behavior* (Philadelphia: J. B. Lippincott, 1973), p. 195, table 6, cols. 5-6. Age ratio: U.S. Bureau of the Census, *Current Population Reports*, series P-25, as follows: 1940-50—no. 98, p. 115; 1955—no. 265, p. 25; 1960—no. 286, p. 42, series C; 1965—no. 519, p. 20; 1970-75—no. 614, pp. 11-16.

APPENDIX TABLE 4.1

*Labor Force Participation Rates of Females
in Specified Age Group, 1890-1975*

Year	(1) 20-24	(2) 25-34	(3) 45-54	(4) 55-64
1890	32.7	18.8	17.4	14.8
1900	34.2	21.7	19.1	16.0
1920	40.0	25.7	22.8	17.7
1930	44.3	29.1	24.6	18.7
1940	48.1	35.3	27.3	20.0
1950	46.1	34.0	38.0	27.0
1955	46.0	34.9	43.8	32.5
1960	46.2	36.0	49.8	37.2
1965	50.0	38.6	50.9	41.1
1970	57.8	45.0	54.4	43.0
1975	64.3	54.6	54.6	41.0

Source: 1950-1975—*Employment and Training
Report of the President*; 1890-1940—extrapola-
tion of 1950 value in ibid. by percentage point
change shown in Gertrude Bancroft, *The
American Labor Force* (New York: Wiley,
1958), p. 207.

APPENDIX TABLE 5.1

Percentage of First Marriages Ending in Divorce, and Children Ever
Born for Cohorts Married 1910-14 to 1955-59

Period of first marriage	(1) Actual	(2) Trend value	(3) Deviation from trend (1)-(2)	(4) Children ever born
	First marriages ending in divorce, percent			
1910-14	14.56	15.22	−.66	3.337
1915-19	16.98	16.78	.20	2.968
1920-24	18.60	18.43	.17	2.637
1925-29	21.18	20.18	1.00	2.449
1930-34	24.30	22.01	2.29	2.417
1935-39	25.56	23.92	1.64	2.497
1940-44	26.80	25.91	.89	2.605
1945-49	25.88	27.97	−2.09	2.838
1950-54	28.42	30.08	−1.66	3.133
1955-59	32.16	32.24	−.08	3.076

Source: Samuel H. Preston and John McDonald, "The Incidence of
Divorce Within Cohorts of American Marriages Contracted Since the
Civil War," Demography 16 (1) (February 1979). Col. 1 is five-year
averages of pp. 10-11, table 2; col. 4 for 1915-59; and table 1, col. 3,
p. 4, for 1910-14. Col. 2 is five-year averages of estimates from the
trend equation given on page 13. Col. 4 is from table 3, col. 2, p. 16,
except for the 1955-59 and 1950-54 figures. For these years, the average number of children ever born in the Preston-McDonald table was
inflated by 12.5 percent and 5.7 percent respectively as recommended
by them to allow for the additional childbearing expected in these two
cohorts beyond 1970.

APPENDIX TABLE 5.2

*Percentage of Premaritally
Conceived First Births
Legitimated by Marriage,
Females 15-19[a] 1939-42 to
1975-78*

Period of first birth	Percent
1939-42	67.3
1943-46	59.8
1947-50	60.3
1951-54	60.9
1955-58	75.7
1959-62	71.3
1963-66	71.1
1967-70	70.9
1971-74	64.7
1975-78	58.3[b]

[a] Age at first birth.
[b] Base exludes permarital births to women who were never married and fifteen to seventeen years old at the time of the survey.

Source: Martin O'Connell and Maurice J. Moore, "The Legitimacy Status of First Births: A Retrospective Study, 1939-42 to 1975-78," *Family Planning Perspectives* vol. 12, no. 1 (Jan./Feb. 1980).

APPENDIX TABLE 6.1

Subjective Feelings of Well-Being, by Age, 1957 and 1976

	(1) (2) Happiness: Percent "very happy"		(3) (4) Worrying: Percent saying "always"; "a lot"		(5) (6) Psychological anxiety: Percent "very high"	
Age	1957	1976	1957	1976	1957	1976
21-29	39	31	32	51	10	16
30-39	51	34	32	53	14	17
40-49	32	32	34	47	14	18
50-59	35	31	35	39	18	16
60-64	26	25	33	42	18	18
65+	25	29	36	39	23	17

Source: Joseph Veroff, "General Feelings of Well Being Over a Generation: 1957-1976," unpublished paper presented to American Psychological Association, September 1, 1978, tables 1-3. The specific questions asked were as follows:

Columns 1 and 2. "Taking things altogether, how would you say things are these days—would you say they're very happy, pretty happy, or not too happy these days?"

Columns 3 and 4. "Everyone has some things he worries about more or less. What kinds of things do you worry about? Do you worry about such things a lot or not very much?"

Columns 5 and 6. Sum of the following five: (These questions all have a range of possible answers including (1) nearly all the time; (2) pretty often; (3) not very much; and (4) never, and are weighted respectively 4, 3, 2, 1 within each question for this scale.)

a. "Do you ever have any trouble getting to sleep or staying asleep?"
b. "Have you ever been bothered by nervousness, feeling fidgety and tense?"
c. "Are you ever troubled by headaches or pains in the head?"
d. "Do you have loss of appetite?"
e. "How often are you bothered by having an upset stomach?"

APPENDIX TABLE 6.2

Death Rates Due to Homicide and Suicide,
Males Aged 15-24; and Death Rate due to
Suicide, Males Aged 45-54, 1946-1977
(per 100,000)

Year	(1) Males 15-24 Homicide	(2) Males 15-24 Suicide	(3) Males 45-54 Suicide
1946	12.2	7.4	30.3
1947	11.9	6.6	30.5
1948	11.3	6.6	31.5
1949	10.1	6.7	31.6
1950	9.6	6.5	32.0
1951	9.0	6.5	28.5
1952	9.9	6.5	27.9
1953	9.6	6.5	29.2
1954	9.3	6.7	31.0
1955	8.5	6.3	29.7
1956	9.4	6.3	28.3
1957	8.9	6.4	28.6
1958	8.8	7.4	32.1
1959	9.0	7.7	31.0
1960	9.1	8.2	31.6
1961	8.8	7.9	31.0
1962	9.0	8.5	30.9
1963	9.0	9.0	30.7
1964	9.9	9.2	29.9
1965	10.7	9.4	29.1
1966	12.0	9.7	28.5
1967	14.3	10.5	27.9
1968	16.4	10.9	27.3
1969	18.0	12.2	27.2
1970	19.0	13.5	27.9
1971	20.6	14.0	26.8
1972	21.5	15.7	28.0
1973	20.7	17.0	26.9
1974	22.1	17.1	26.6
1975	21.2	18.9	27.9
1976	19.1	18.5	26.2
1977	19.4	21.8	25.6

Sources: 1946-60: National Center for Health Statistics, U.S. Department of Health, Education, and Welfare, *Vital Statistics Rates in the United States: 1940-1960* (Washington, D.C.: U.S. Government Printing Office, 1968), table 63.

1961-75: National Center for Health Statistics, U.S. Department of Health, Education, and Welfare, *Vital Statistics of the United States*, Annual Yearbook (various issues), vol. II *Mortality*, part A.

1976-77: Unpublished data, Mortality Division of National Center for Health Statistics.

APPENDIX TABLE 6.3

*Percentage of Males Aged 18-24 Answering
Affirmatively on Two Measures of Political
Alienation, 1952-76*

Year	(1) Agrees that Government is Too Complicated	(2) Agrees that He Has No Say in Government
1952	71.4	26.1
1956	62.2	23.4
1964	65.7	24.5
1966	59.5	25.0
1970	72.3	34.9
1972	71.9	41.0
1974	69.9	35.9
1976	72.0	39.9

Source: Data from the University of Michigan Survey Research Center's presidential election surveys. The specific statements were:

Column 1: "Sometimes politics and government seem so complicated that a person like me can't really understand what's going on."
Column 2: "People like me don't have any say about what the government does."

APPENDIX TABLE 7.1

Comparative Change in Labor Force and Unemployment for Four Sex-Age Groups,
1956-74[a]

	(1)	(2)	(3)	(4)	(5)
				Unemployment Rate	
	Percentage increase in civilian labor force, 1956-74	1956	1974	Change, percentage points (3) − (2)	Change, percentage (4) ÷ (2)
Females under 25	127	8.3	12.3	+4.0	48
Males under 25	104	8.6	11.4	+2.8	32
Females 25 and over	52	4.0	4.6	+0.6	15
Males 25 and over	10	3.1	3.0	−0.1	−3

[a] The comparison is for two dates when the economy is estimated to have been close to a full-employment equilibrium. See Michael L. Wachter and Jeffrey M. Perloff, "A Production Function-Nonaccelerating Inflation Approach to Potential Output: Is Measured Potential Output Too High?" *Carnegie-Rochester Conference Series on Public Policy* (a supplement to the *Journal of Monetary Economics*), vol. 10, 1979.

Source: Bureau of Labor Statistics, *Handbook of Labor Statistics, 1968*, U.S. Department of Labor, Bulletin No. 1600 (Washington, D.C.: U.S. Government Printing Office, 1968), p. 27, p. 96.
Bureau of Labor Statistics, *Handbook of Labor Statistics, 1977*, U.S. Department of Labor, Bulletin No. 1966 (Washington, D.C.: U.S. Government Printing Office, 1977), p. 25, p. 110.

APPENDIX TABLE 7.2

Growth Rate of Consumer Price Index, 1940-79

Period	Growth Rate of Consumer Price Index
1940-50	5.6
1950-55	2.2
1955-60	2.1
1960-65	1.3
1965-70	4.3
1970-75	6.7
1975-79[a]	7.6

[a]May to May.

Source: 1940-65 – Bureau of Labor Statistics, *Handbook of Labor Statistics, 1968*, U.S. Department of Labor, Bulletin no. 1600 (Washington, D.C.: U.S. Government Printing Office, 1968), p. 256. 1965-75 – Bureau of Labor Statistics, *Handbook of Labor Statistics, 1977*, U.S. Department of Labor, Bulletin no. 1966 (Washington, D.C.: U.S. Government Printing Office, 1977), p. 239. May 1975 – Bureau of Economic Analysis, *Survey of Current Business*, U.S. Department of Commerce, vol. 55, no. 9 (Washington, D.C.: U.S. Government Printing Office, September 1975), p. S-8. May 1979 – Bureau of Economic Analysis, *Business Conditions Digest*, U.S. Department of Commerce, vol. 19, no. 6 (Washington, D.C.: U.S. Government Printing Office, 1979), p. 84.

APPENDIX TABLE 8.1

Average Growth Rate of Population by Component of
Change, 1870-1978
(per 1,000 per year)

Period	Rate of Population Growth	Net Immigration Rate	Birth Rate	Death Rate
1870-75	25.5	6.7	40.8	21.8
1875-80	18.3	3.4	38.8	23.8
1880-85	25.4	10.1	36.9	21.0
1885-90	19.9	5.8	35.3	20.6
1890-95	20.1	4.5	34.3	19.5
1895-1900	16.3	2.8	31.6	18.8
1900-05	18.5	6.0	30.0	17.6
1905-10	19.8	6.9	29.6	16.6
1910-15	17.5	5.3	27.5	14.7
1915-20	10.5	1.1	26.1	16.2
1920-25	16.9	3.6	25.0	11.3
1925-30	12.5	2.0	21.5	10.6
1930-35	7.0	−0.4	18.3	11.0
1935-40	7.2	0.2	18.3	11.3
1940-45	10.6	0.5	21.2	10.9
1945-50	15.6	1.3	24.5	9.9
1950-55	16.9	1.2	25.2	9.6
1950-55	17.2	1.8	24.8	8.9
1955-60	17.0	1.8	24.6	9.4
1960-65	14.5	1.9	22.2	9.4
1965-70	10.6	2.2	18.1	9.5
1970-75	8.3	1.7	15.8	9.5
1975-78	7.6	1.6	14.9	8.8

Sources: 1870-1955 — Except as noted subsequently, the basic source was Simon Kuznets, "Long Swings in the Growth of Population and in Related Economic Variables," *Proceedings of the American Philosophical Society* 102 (1) (1958), p. 37, table 1, col. 4; p. 39, table 3, col. 5; p. 41, table 5, col. 7; and p. 43, table 6, col. 5 (underlying unrounded quinquennial estimates were used). However, for net migration, 1910-40 data were from Simon Kuznets and E. Rubin, *Immigration and the Foreign Born* (New York: National Bureau of Economic Research, 1954), pp. 95-96, table B-1. For 1940-55 — estimates for all series were revised somewhat, the chief differences from the original source being inclusion of armed forces deaths overseas and keeping the scope of the net migration estimate the same as for the pre-1940 period.

The sum of the components does not exactly equal total increase because net migration refers to alien arrivals less departures and thus includes some nonmigratory movements. Pure migration estimates are not available prior to 1910.

1950-78: U.S. Bureau of the Census, *Current Population Reports*, series P-25, no. 706 (September 1977), p. 7; U.S. Department of Health, Education, and Welfare, *Monthly Vital Statistics Report: Births, Marriage, Divorces and Deaths for June 1978* (12 September 1978). U.S. Department of Commerce, Bureau of the Census, *Estimate of the Population of the United States to July 1, 1978*, Series P-25, No. 729 (August 1978).

NOTES

Preface

1. Richard A. Easterlin, "The American Baby Boom in Historical Perspective" (New York: National Bureau of Economic Research, Occasional Paper 79, 1962)), p. 32; and *Population, Labor Force, and Long Swings in Economic Growth* (New York: National Bureau of Economic Growth, 1968).

Chapter 1

1. Alice S. Rossi, "Transition to Parenthood," in Jerold Heiss, ed., *Family Roles and Interaction: An Anthology,* 2nd ed. (Chicago: Rand McNally, 1976), p. 276. See also Judith Blake, "Coercive Pronatalism and American Population Policy," in Robert Parke, Jr., and Charles F. Westoff, eds., *Commission on Population Growth and the American Future, Research Reports, vol. VI, Aspects of Population Growth Policy* (Washington, D.C.: U.S. Government Printing Office, 1972).

2. See Janet Salzman Chafetz, *Masculine, Feminine, or Human: An Overview of the Sociology of the Gender Roles,* 2nd ed. (Itasca, Ill.: F. E. Peacock, 1978), chap. 3. Also Letty Cottin Pogrebin, "Down with Sexist Upbringing," in Heiss, *Family Roles and Interaction,* pp. 321–31, and Joan D. Mandle, *Women and Social Change in America* (Princeton: Princeton Book Company, 1979), chap. 3.

3. Karen Oppenheim Mason, "Studying Change in Sex-Role Definitions via Attitude Data," *American Statistical Association: Proceedings of the Social Statistics Section* (1973), pp. 138–39; and Arland Thornton and Deborah Freedman, "Changes in the Sex Role Attitudes of Women, 1962–1977: Evidence from a Panel Study," *American Sociological Review* (October 1979) 44 (5): 832–42.

4. Bureau of the Census, *Fertility of American Women: June 1977,* U.S. Bureau of the Census, *Current Population Reports,* series P–20, no. 325 (Washington, D.C.: U.S. Government Printing Office, September 1978), p. 27.

5. Judith Blake, "Can We Believe Recent Data on Birth Expectations in the United States," *Demography* (February 1974) 11 (1): 36.

6. A. Regula Herzog, Jerald G. Bachman, and Lloyd D. Johnson, "High School Seniors' Preferences for Sharing Work and Family Responsibilities between Husband and Wife," Institute for Social Research, University of Michigan, Ann Arbor, 1978.

Notes

7. Data supplied by the Gallup poll, American Institute of Public Opinion, Princeton, New Jersey.

8. Tabulation of unpublished data from the National Longitudinal Survey of Young Women. The original national sample was of females aged fourteen to twenty-four in 1968; the same women were then surveyed again in 1975 aged twenty-one to thirty-one.

Chapter 2

1. William Wordsworth, *The Prelude.*

2. Denis F. Johnston, "The Aging of the Baby Boom Cohorts," *Statistical Reporter,* nos. 76–79 (Washington, D.C.: U.S. Government Printing Office, March 1976), pp. 161–65. I have taken this from Carl L. Harter, "The 'Good Times' Cohort of the 1930s (Sometimes Less Means More and More Means Less)," *Population Reference Bureau Report* (Washington, D.C.: Population Reference Bureau, Inc., April 1977) 3 (3), which provides a good exposition of the notion of birth cohorts.

3. The chart is confined to year-round full-time workers to get as nearly as possible to salary or pay *rates.*

4. For the details of a test of this, see Richard A. Easterlin, "What Will 1984 Be Like? Socioeconomic Implications of Recent Twists in Age Structure," *Demography* (November 1978) 15 (4): Appendix B, p. 430.

5. The sizable employment and earnings effects of recent shifts in the relative numbers of younger and older workers, and the low elasticity of substitution between them even after controlling for other pertinent factors, such as education, have been demonstrated in a number of recent studies. *On earnings,* see Richard B. Freeman, "The Effect of Demographic Factors on Age-Earnings Profiles," *The Journal of Human Resources,* 14 (3): 289–318; Ronald Lee, "Causes and Consequences of Age Structure Fluctuations: The Easterlin Hypothesis," in International Union for the Scientific Study of Popuulation, *Economic and Demographic Change: Issues for the 1980's, Helsinki, 1978* (Liege, Belgium: International Union for the Scientific Study of Population, 1979), pp. 405–18; and "Economic Aspects of Age Structure—Introductory Statement," in International Union for the Scientific Study of Population, *Economic and Demographic Change: Issues for the 1980's, Helsinki, 1978* (Liege, Belgium: International Union for the Scientific Study of Population, 1979), pp. 399–403; Finis Welch, "Effects of Cohort Size on Earnings: The Baby Boom Babies' Financial Bust," *Journal of Political Economy* (October 1979) 87 (1): 65–74; and Michael L. Wachter, "Intermediate Swings in Labor-Force Participation," *Brookings Papers on Economic Activity* (1977) 2: 545–76. *On unemployment,* see Joseph M. Anderson, "Population Change and the American Labor Market: 1950–2000," in U.S. House of Representatives, *Consequences of Changing U.S. Population: Baby Boom and Bust,* Hearings Before the Select Committee on Population, Ninety-fifth Congress, second session, vol. II (Washington, D.C.: U.S. Government Printing Office, 23, 25 May and 1, 2 July 1978), pp. 781–804; and "An Economic-Demographic Model of the United States Labor Market," Ph.D. dissertation, Harvard Uni-

Notes

versity, 1977; and Michael L. Wachter, "The Changing Cyclical Responsiveness of Wage Inflation," *Brookings Papers on Economic Activity* (1976) 1: 115–59. An analysis of post-World War II movements in the ratio of earnings by age appears in Valerie Kincade Oppenheimer, *Work and the Family: A Reference Group Analysis* (forthcoming).

6. John E. Buckley, "Pay Differences between Men and Women in the Same Job," *Monthly Labor Review* (November 1971) 94 (11): 36–39.

7. Cf. Gertrude Bancroft, *The American Labor Force: Its Growth and Changing Composition* (New York: John Wiley, 1958), p. 82, on the labor market situation of young women in the 1950s: "The abundance of white-collar jobs which did not require extensive training attracted an increasing proportion of young women and made it possible for them to avoid the lower-paid or less desirable types of work."

8. Freeman and Welch also conclude that the cohort-size effect persists over the life cycle, although noting reasons why it may be attenuated. See Freeman, "The Effect of Demographic Factors," pp. 32–33, and Welch, "Effects of Cohort Size." For a good summary review of the literature on cohorts and aging, see Matilda White Riley, "Age and Aging: From Theory Generation to Theory Testing," paper presented at the annual meeting of the American Sociological Association, Boston, 1979.

9. H. Scott Gordon, "On Being Demographically Lucky: The Optimum Time to Be Born," (presidential address to the Western Economic Association, annual meeting, Anaheim, California, June 1977), p. 4.

10. Ibid., pp. 7–8.

11. The argument in this section is developed fully in Richard A. Easterlin, *Population, Labor Force, and Long Swings in Economic Growth* (New York: National Bureau of Economic Research, 1968), chaps. 2 and 3.

12. Ibid., pp. 32, 44.

13. Stanley Lebergott, *Manpower in Economic Growth: The American Record since 1800* (New York: McGraw-Hill, 1964), p. 187.

14. For example, in the peak five-year period of the last major immigration boom occurring under unrestricted immigration, 1905–10, the average annual rate of immigration was almost 7.0 per 1,000 total population. In 1970–76, the corresponding figure was 1.7; a reasonable allowance for illegal immigration would perhaps double this figure. See Richard A. Easterlin, Michael L. Wachter, and Susan M. Wachter, "The Changing Impact of Population Swings on the American Economy," *Proceedings of the American Philosophical Society* (June 1978) 122 (3): 119–30; and *Legal and Illegal Immigration to the United States*, report prepared by the Select Committee on Population, U.S. House of Representatives (Washington, D.C.: U.S. Government Printing Office, December 1978).

Chapter 3

1. Charles F. Westoff, "Some Speculations on the Future of Marriage and Fertility," *Family Planning Perspectives* (March/April 1978) 10 (2): 79–80. See also Charles F. Westoff, "Marriage and Fertility in the

Notes

Developed Countries," *Scientific American* (December 1978) 239 (6); and "The Predictability of Fertility in Developed Countries," *Population Bulletin of the United Nations* (1978) no. 11 (New York: United Nations, Department of International Economic and Social Affairs, 1978).

2. Ibid., p. 82.

3. "Margaret Mead Predicts Lower Birth Rate in U.S.," *The New York Times,* 30 January 1978, p. 10.

4. "The Population Forecasts of the Scripps Foundation," *Population Index* (1948) 14 (3): 190–95.

5. U.S. Department of Labor, *Career Thresholds,* Volume I, Manpower Research Monograph no. 16 (Washington, D.C.: U.S. Government Printing Office, 1970), p. 122.

6. This is strictly an expository convenience. Evidence suggests that the wife's father's characteristics are important. See Eileen Crimmins-Gardner and Phyllis Ewer, "Relative Status and Fertility," in Julian L. Simon, ed., *Research in Population Economics: An Annual Compilation of Research,* Volume I (Greenwich, Conn.: JAI Press, Inc., 1978); and Lolagene C. Coombs and Zena Zumeta, "Correlates of Marital Dissolution in a Prospective Fertility Study: A Research Note," *Social Problems* (Summer 1970) 18 (1): 96.

7. Lee Rainwater, *What Money Buys* (New York: Basic Books, 1974).

8. Richard A. Easterlin, *Population, Labor Force, and Long Swings in Economic Growth* (New York: National Bureau of Economic Research, 1968), p. 241.

9. See U.S. Department of Commerce, Bureau of the Census, *Current Population Reports,* series P–20, "Marital Status and Living Arrangements," table 1, various years.

10. Ronald R. Rindfuss and James A. Sweet, *Postwar Fertility Trends and Differentials in the United States* (New York: Academic Press, 1977).

11. Glen H. Elder, Jr., *Children of the Great Depression: Social Change in Life Experience* (Chicago: University of Chicago Press, 1974), p. 196.

12. Selma Taffel, "Trends in Fertility in the United States," National Center for Health Statistics, *Vital and Health Statistics,* U.S. Department of Health, Education, and Welfare, series 21, no. 28 (Hyattsville, Md., September 1977).

13. Judith Blake, "Can We Believe Recent Data on Birth Expectations in the United States," *Demography* (February 1974) 11 (1): 25–44.

14. Westoff, "Marriage and Fertility in the Developed Countries," p. 53.

15. Charles F. Westoff and Norman B. Ryder, *The Contraceptive Revolution* (Princeton: Princeton University Press, 1977), p. 340.

16. See Pascal K. Whelpton, Arthur A. Campbell, and John E. Patterson, *Fertility and Family Planning in the United States* (Princeton, N.J.: Princeton University Press, 1966), chap. 5.

17. Jeanne Clare Ridley, "A Study of Low Fertility Cohorts in the United States," paper in progress, Kennedy Institute, Center for Population Research.

18. United Nations, Department of Economic and Social Affairs, *Fertility and Family Planning in Europe Around 1970: A Comparative Study of Twelve National Surveys,* Population Studies no. 85 (New York: United Nations, 1976), p. 151.

Notes

19. Richard A. Easterlin, "Relative Economic Status and the American Fertility Swing," in Eleanor Bernert Sheldon, ed., *Family Economic Behavior: Problems and Prospects* (Philadelphia: J. B. Lippincott Company, 1972), p. 187.

20. Norman B. Ryder and Charles F. Westoff, *Reproduction in the United States 1965* (Princeton, N.J.: Princeton University Press, 1971), p. 152.

21. Norman B. Ryder, "A Model of Fertility by Planning Status," *Demography*, (November 1978) 15 (4): 433–58.

22. Blake, "Can We Believe Recent Data," p. 27.

23. See "Sloppier Contraception, Not More Wanted Births, Caused 1950s Baby Boom, NSF Researcher Says," *Family Planning Perspectives* (November/December 1978) 10 (6): 369.

24. Jeanne Clare Ridley, "The Changing Position of Women: Education, Labor Force Participation and Fertility," in Fogarty International Center, *The Family in Transition* (Washington, D.C.: U.S. Government Printing Office, 1971).

25. "Money Need, Social Change Combine to Cut Apron Strings," *The New York Times,* 7 May 1978, section IV, p. 2.

26. See "Kindergarten and Elementary School Teachers," Bureau of Labor Statistics, *Occupational Outlook Handbook, 1949,* U.S. Department of Labor, Bulletin no. 940 (Washington, D.C.: U.S. Government Printing Office, 1949), p. 38; "Kindergarten and Elementary School Teachers," Bureau of Labor Statistics, *Occupational Outlook Handbook, 1951,* U.S. Department of Labor, Bulletin no. 998 (Washington, D.C.: U.S. Government Printing Office, 1951), p. 48; and "Registered Professional Nurses," *Occupational Outlook Handbook, 1949,* p. 49.

Chapter 4

1. See chap. 3, note 25. Since 1940 sex polarization of occupations has increased. See Hilda Kahne and Andrew I. Kohen, "Economic Perspectives on the Roles of Women in the American Economy," *Journal of Economic Literature* (December 1975) 13 (4): 1274.

2. R. D. Barron and G. M. Norris, "Sexual Divisions and the Dual Labour Market," in Diana Barker and Sheila Allen, eds., *Dependence and Exploitation in Work and Marriage* (London: Longman, 1976).

3. Kahne and Kohen, "Economic Perspectives," p. 1274.

4. Deborah Pisetzner Klein, "Women in the Labor Force: The Middle Years," *Monthly Labor Review* (November 1975) 98 (11): 13. If occupational level and income are held constant, males and females do not differ significantly in turnover rates (Janet Salzman Chafetz, *Masculine, Feminine, or Human: An Overview of the Sociology of the Gender Roles,* 2nd ed. [Itasca, Ill.: F. E. Peacock, 1978, pp. 132–33]).

5. See chap. 1, note 8.

6. Bureau of Labor Statistics, *U.S. Working Women: A Databook,* U.S. Department of Labor, Bulletin 1977 (Washington, D.C.: U.S. Government Printing Office, 1977), p. 9.

Notes

7. Ruth Gilbert Shaeffer and Edith F. Lynton, *Corporate Experiences in Improving Women's Job Opportunities* (New York: The Conference Board, 1979), p. 11.

8. Bureau of the Census, *Current Population Reports,* series P–60, no. 118, pp. 180, 182.

9. Charles F. Westoff, "Some Speculations on the Future of Marriage and Fertility," *Family Planning Perspectives* (March/April 1978) 10 (2): 81.

10. Data for 1940–U.S. Bureau of the Census, *Census of Population: 1950, Volume II, Characteristics of the Population, Part I, United States Summary* (Washington, D.C.: U.S. Government Printing Office, 1953), table 115, p. 1–238; data for 1950–*ibid.,* table 115, p. 1–236; 1960–U.S. Bureau of the Census, *Census of Population: 1960, Volume 1, Characteristics of the Population, Part I, United States Summary* (Washington, D.C.: U.S. Government Printing Office, 1964), table 173, p. 1–405; 1970–U.S. Bureau of the Census, *Census of Population: 1970, Detailed Characteristics Final Report, PC(1)–D1, U.S. Summary* (Washington, D.C.: U.S. Government Printing Office, 1973), table 199, p. 1–627; 1977–U.S. Bureau of the Census, "Educational Attainment in the United States: March 1977 and 1976," *Current Population Reports,* series P–20, no. 314 (Washington, D.C.: U.S. Government Printing Office, 1977), table 1, p. 7.

11. See Richard A. Easterlin, "Fertility and Female Labor Force Participation in the United States: Recent Changes and Future Prospects," in International Union for the Scientific Study of Population, *Economic and Demographic Change: Issues for the 1980's, Helsinki, 1978* (Liege, Belgium: International Union for the Scientific Study of Population, 1979), pp. 71–86. For similar views, see Michael L. Wachter, "A Labor Supply Model for Secondary Workers," *Review of Economics and Statistics* (May 1972) 54 (2): 141–51, and Joan D. Mandle, *Women and Social Change in America* (Princeton, N.J.: Princeton Book Company, 1979), pp. 115–17.

12. Gertrude Bancroft, *The American Labor Force: Its Growth and Changing Composition* (New York: John Wiley, 1958), p. 80; see also p. 30.

13. Ibid., p. 132.

14. Michael L. Wachter, "Intermediate Swings in Labor-Force Participation," *Brookings Papers on Economic Activity* 2 (1977).

15. William P. Butz and Michael P. Ward, *The Emergence of Countercyclical U.S. Fertility* (Santa Monica, Calif.: The Rand Corporation, June 1977), p. 18.

16. William P. Butz and Michael P. Ward, *Countercyclical U.S. Fertility and Its Implications* (Santa Monica, Calif.: The Rand Corporation, November 1978), pp. 9–10.

17. Richard A. Easterlin, "What Will 1984 Be Like? Socioeconomic Implications of Recent Twists in Age Structure," *Demography* (November 1978) 15 (4): 431–32.

18. Robert Zajonc, "Thinking: Cognitive Organization and Processes," *International Encyclopedia of the Social Sciences,* vol. 15 (New York: Macmillan, 1968), pp. 618–21.

19. Karen Oppenheim Mason, John L. Czajka, and Sara Arber, "Change in U.S. Women's Sex-Role Attitudes, 1964–1974," *American Sociological Review* (August 1976) 41 (4): 575.

Notes

Chapter 5

1. William J. Goode, "Family Disorganization," in Robert K. Merton and Robert Nisbet, eds., *Contemporary Social Problems* (New York: Harcourt Brace, 1971), pp. 467–70.
2. Paul C. Glick and Arthur J. Norton, "Marrying, Divorcing, and Living Together in the U.S. Today," *Population Bulletin* (Washington, D.C.: Population Reference Bureau, Inc., October 1977) 32 (5): 3.
3. Ibid.
4. For fuller discussion of developments in divorce, see the comprehensive empirical study by Hugh Carter and Paul C. Glick, *Marriage and Divorce: A Social and Economic Study* (Cambridge: Harvard University Press, 1976). Valuable recent surveys of illegitimacy are Catherine S. Chilman, *Adolescent Sexuality in a Changing American Society: Social and Psychological Perspectives,* U.S. Department of Health, Education, and Welfare, Public Health Service, National Institute of Health (Washington, D.C.: U.S. Government Printing Office, 1979), DHEW Publication Number (NIH) 79–1426; Phillips Cutright, "Illegitimacy in the United States," in Charles F. Westoff and Robert Parke, Jr., eds., Commission on Population Growth and the American Future, *Research Reports, Volume I, Demographic and Social Aspects of Population Growth* (Washington, D.C.: U.S. Government Printing Office, 1972), pp. 376–438; and Select Committee on Population, *Fertility and Contraception in America: Adolescent and Pre-Adolescent Pregnancy,* U.S. Senate, Ninety-Fifth Congress, Second Session, Hearings, 28 February, 1, 2 March 1978 (Washington, D.C.: U.S. Government Printing Office, 1978).
5. The view here is consistent with sociologist John Scanzoni's theory of marital dissolution, in which the emphasis is on the spouses' performance of "role obligations." See *Opportunity and the Family* (New York: Free Press, 1970). Except for recent work by Becker, Landes, and Michael, which ignores or minimizes the type of relative income mechanism stressed here and by sociologists, economists have almost wholly neglected the analysis of marital dissolution. See Gary S. Becker, Elisabeth M. Landes, and Robert T. Michael, "An Economic Analysis of Marital Instability," *Journal of Political Economy* (December 1977) 85 (6): 1141–88 and Robert T. Michael, "Two Papers on the Recent Rise in U.S. Divorce Rates," Working Paper No. 202, Center for Economic Analysis of Human Behavior and Social Institutions, National Bureau of Economic Research, Inc., Stanford, Calif., September 1977.
6. As quoted in Arlene Skolnick, *The Intimate Environment: Exploring Marriage and the Family,* 2nd ed. (Boston: Little, Brown, 1978), p. 217.
7. National Center for Health Statistics, "Divorces: Analysis of Changes, United States, 1969," *Vital and Health Statistics,* U.S. Department of Health, Education, and Welfare, series 21, no. 22 (Washington, D.C.: U.S. Government Printing Office, April 1973); and Andrew Cherlin, "The Effect of Children on Marital Dissolution," *Demography* (August 1977) 14 (3): 265–72.
8. Gerald C. Wright, Jr., and Dorothy M. Stetson, "The Impact of No-Fault Divorce Law Reform on Divorce in American States," *Journal of Marriage and the Family* (August 1978) 40 (3): 575–80.

Notes

9. Lolagene C. Coombs and Zena Zumeta, "Correlates of Marital Dissolution in a Prospective Fertility Study: A Research Note," *Social Problems* (Summer 1970) 18 (1): 92–101.

10. Ibid., p. 96.

11. Samuel H. Preston and John McDonald, "The Incidence of Divorce Within Cohorts of American Marriages Contracted Since the Civil War," *Demography* (February 1979) 16 (1): 1–25.

12. Ibid., p. 20.

13. John F. Kantner and Melvin Zelnik, "Sexual Experiences of Young Unmarried Women in the United States," *Family Planning Perspectives* (October 1972) 4 (4): 11. See also Melvin Zelnik and John F. Kantner, "Sexual and Contraceptive Experience of Young Unmarried Women in the United States, 1976 and 1971," *Family Planning Perspectives* (March/April 1977) 9 (2): 56, 61.

14. Martin O'Connell and Maurice J. Moore, "The Legitimacy Status of First Births, 1939–78," *Family Planning Perspectives* (January/February 1980) 12 (1).

15. Martin O'Connell, "A Cohort Analysis of Teenage Fertility in the U.S. Since the Depresion," paper presented at the Population Association of America, annual meeting, Atlanta, Ga., April 1978, table 1. Premarital conceptions are estimated, not from personal reports, but from a comparison of dates of marriage and first births. Births occurring up to and including seven months after marriage are counted as premarital conceptions.

16. O'Connell and Moore, "Legitimacy Status."

17. The percentage impact on the illegitimacy rate is approximately equal to the percentage change in the proportion of births *not* legitimated by marriage $\left(\frac{15}{100-75} \right) \times 100 = 60$ percent. The source for illegitimacy rates is National Center for Health Statistics, "Teenagers: Marriages, Divorces, Parenthood, and Mortality," *Vital and Health Statistics,* U.S. Department of Health, Education, and Welfare, series 21, no. 23 (Washington, D.C.: U.S. Government Printing Office, August 1973), p. 32; and National Center for Health Statistics, "Final Natality Statistics, 1976," *Monthly Vital Statistics Report,* U.S. Department of Health, Education, and Welfare, vol. 26, no. 12, Supplement (Hyattsville, Maryland, 29 March 1978).

Chapter 6

1. My thinking on these subjects has benefited particularly from Martin O'Connell's doctoral dissertation. See Martin O'Connell, "The Effect of Changing Age Distributions on Fertility and Suicide in Developed Countries," unpublished Ph.D. dissertation, University of Pennsylvania, 1975. An unpublished paper by James Taylor develops an argument similar to the present. See James B. Taylor, Martha Carithers, and Lolafaye Coyne, "Explaining Historical Trends in Collective Deviance and Psychopathology," unpublished paper presented to Mid-America Psychosocial Studies Group, Topeka, Kan., 21 January 1976. Waldron and Eyer document a number of the developments since 1960 discussed here but do not link

Notes

them to generation size. See Ingrid Waldron and Joseph Eyer, "Socioeconomic Causes of the Recent Rise in Death Rates for 15–24 Year Olds," *Social Science and Medicine* (1975) 9: 383–96.

2. The view presented here draws significantly on sociological theory. See Robert K. Merton, *Social Theory and Social Structure*, 1968 enlarged ed. (New York: The Free Press, 1968); Andrew F. Henry and James F. Short, Jr., *Suicide and Homicide* (New York: The Free Press of Glencoe, 1954); Albert K. Cohen and James F. Short, Jr., "Crime and Juvenile Delinquency" in Robert K. Merton and Robert Nisbet, *Contemporary Social Problems*, 3rd ed. (New York: Harcourt Brace Jovanovich, 1971); Jack P. Gibbs, "Suicide," in ibid. Few economists have dealt with questions such as crime and suicide, and those who have usually disregard the sociologists' stress on aspirations. *On crime*, see G. S. Becker, "Crime and Punishment: An Economic Approach," *Journal of Political Economy*, vol. 26 (March/April 1968); Kenneth I. Wolpin, "An Economic Analysis of Crime and Punishment in England and Wales, 1894–1967," *Journal of Political Economy* (1978) 86 (5): 815–40. *On suicide*, see Daniel S. Hamermesh and N. Soss, "An Economic Theory of Suicide," *Journal of Political Economy* (January/February 1974) 82 (1): 83–98. A good elementary exposition of this work appears in Robert Crouch, *Human Behavior, An Economic Approach* (North Scituate, Mass.: Duxbury Press, 1979). An exception among economists in adopting a sociological view is the work of Julian Simon and his associates on suicide. See Julian L. Simon, "The Effect of Income on the Suicide Rate: A Paradox Resolved," *The American Journal of Sociology* (November 1968) 74 (3): 302–3; Carl B. Barnes, "The Partial Effect of Income on Suicide is Always Negative," *The American Journal of Sociology* (May 1975) 80: 1454–60; Julian L. Simon, "Response to Barnes' Comment," *The American Journal of Sociology* (May 1975) 80: 1460–62. Recently, there have been attempts by sociologists to employ econometric models in the study of these subjects, but the theory employed tends to follow the lead of economists rather than of sociologists. See James Alan Fox, "An Econometric Analysis of Crime Data," unpublished Ph.D. dissertation, University of Pennsylvania, 1976; Stephen S. Brier and Stephen E. Feinberg, "Recent Econometric Modelling of Crime and Punishment: Support for the Deterrence Hypothesis?" Technical Report No. TR–307, University of Minnesota, Department of Applied Statistics, 7 January 1978; Kenneth C. Land and Marcus Felson, "A Dynamic Macro Social Indicator Model of Changes in Marriage, Family, and Population in the United States: 1947–1974," *Social Science Research* (1977) 6: 328–62; Gideon Vigderhous, "Box Jenkins Models of Trends in Suicide," in C. F. Schuessler, ed., *Social Methodology* (San Francisco: Jossey-Bass, 1977), pp. 20–51.

3. Joseph Veroff, "General Feelings of Well Being Over a Generation: 1957–1976," unpublished paper presented to American Psychological Association, 1 September 1978, tables 1–3.

4. Some of these studies are reported in Richard A. Easterlin, "Does Economic Growth Improve the Human Lot?" in *Nations and Households in Economic Growth: Essays in Honor of Moses Abramovitz*, eds. Paul A. David and Melvin W. Reder (New York: Academic Press, Inc., 1974); and John P. Robinson and Phillip R. Shaver, *Measures of Social Psychological Attitudes* (Ann Arbor: Institute for Social Research, August 1969).

5. James Q. Wilson, "Crime Amidst Plenty: The Paradox of the Sixties," in *Thinking about Crime* (New York: Basic Books, Inc., 1975).

6. Charles F. Welford, "Age Composition and the Increase in Recorded Crime," *Criminology* (May 1973) 61: 61–71. See also Wilson, *Thinking about Crime,* p. 17.

7. See also A. Joan Klebba, "Homicide Trends in the United States, 1900–74," *Public Health Reports* (May–June 1975) 90 (3): 195–204; U.S. Department of Health, Education, and Welfare, *Homicide in the United States 1950–1964,* National Center for Health Statistics, series 20, no. 6 (Washington, D.C.: U.S. Government Printing Office, 1967). The series in figures 6.1 start with 1946 rather than 1940 in order to avoid the disturbances associated with World War II.

8. This point is supported by the FBI's data on murder arrests, available since the early 1960s. For males aged fifteen to twenty-four, both the size of the arrest rate and the increase between 1964 and 1976 are very close to the values shown by the data on victims. The arrest data are from Federal Bureau of Investigation, *Uniform Crime Reports,* 1960 through 1976; the population data used in computing arrest rates are from Bureau of the Census, *Current Population Reports,* series P–25.

9. National Center for Health Statistics, *Vital Statistics Rates in the United States: 1940–1960* (Washington, D.C.: 1968), table 63; and *Vital Statistics of the United States* (annual yearbook, 1960–73), vol. 2, *Mortality.* Rates for recent years were obtained directly from the Mortality Division of the National Center for Health Statistics.

10. For the data, see ibid., and U.S. Department of Health, Education, and Welfare, *Suicide in the United States 1950–1964,* National Center for Health Statistics, series 20, No. 5 (Washington, D.C.: U.S. Government Printing Office, 1967). The homicide rate for females was omitted from the previous section, partly because homicide is predominately a male phenomenon and partly because the victimization data for females cannot be taken as indicative of commission of murder.

11. See also U.S. Department of Health, Education, and Welfare, *Suicide in the United States 1950–1964,* National Center for Health Statistics, series 20, no. 5 (Washington, D.C.: U.S. Government Printing Office, 1967), p. 8.

12. See note 8 for reference.

13. O'Connell, "The Effect of Changing Age Distributions."

14. William R. Lyster, "Homicide and Fertility Rates in the United States," *Social Biology* (Winter 1974) 21 (4): 389–92.

15. See O'Connell, "The Effect of Changing Age Distributions."

16. O'Connell, ibid., p. 113, table 3.18.

17. See "Death Rate Is Rising for Nation's Teens," *Washington Post,* 10 November 1979, p. C1.

18. See Ingrid Waldron and Joseph Eyer, "Socioeconomic Causes of the Recent Rise in Death Rates for 15–24 Year Olds," *Social Science and Medicine* (1975) 9: 383–96; also Noel S. Weiss, "Recent Trends in Violent Deaths among Young Adults in the United States," *American Journal of Epidemiology* (1976) 103 (4): 416–22. A positive correlation among city differences in suicide, homicide, and traffic fatalities has been pointed out by Porterfield. See Austin L. Porterfield, "Traffic Fatalities,

Notes

Suicide, and Homicide," *American Sociological Review* (December 1960) 25 (6): 897–901.

19. *Washington Post,* op. cit.
20. Ibid., op. cit.
21. Ibid.
22. Waldron and Eyer, "Socioeconomic Causes," p. 393.
23. The influence of such developments may possibly explain why House and Mason fail to find evidence of significant differential change among age groups like that exhibited by the other measures discussed here. See James S. House and William M. Mason, "Political Alienation in America, 1952–1968," *American Sociological Review* (April 1975) 40 (2): 123–47.
24. I am grateful to sociologist Seymour Martin Lipset for calling my attention to this, and I have drawn here liberally on his valuable survey of the subject. See Seymour Martin Lipset, *Rebellion in the University* (Chicago: University of Chicago Press, 1971), introduction to the Phoenix Edition, especially pp. xxxiii ff.
25. Gabriel Almond, "Youth and Changing Political Culture in America" unpublished paper, Department of Political Science, Stanford University, September 1975, pp. 7–8, as cited in Lipset, *Rebellion in the University.*
26. James W. Kuhn, "The Immense Generation," *The Columbia Forum* 2 (Summer 1973), p. 14, as cited in Lipset, *Rebellion in the University.*

Chapter 7

1. Bureau of Labor Statistics, *Employment and Earnings, January 1979,* vol. 26, no. 1 (Washington, D.C.: U.S. Government Printing Office, 1979), p. 156; Bureau of Labor Statistics, *Handbook of Labor Statistics, 1978,* Bulletin no. 1600 (Washington, D.C.: U.S. Government Printing Office, 1968), p. 256; Bureau of Labor Statistics, *Handbook of Labor Statistics, 1978,* Bulletin no. 1966 (Washington, D.C.: U.S. Government Printing Office, 1977), p. 239.
2. Lawrence R. Klein, "The Supply Side," presidential address delivered at the ninetieth meeting of the American Economic Association, *American Economic Review* (March 1978) 68 (1): 1–7. Sidney Weintraub was one of the first in the United States to stress supply considerations. See S. Weintraub, "A Macroeconomic Approach to the Theory of Wages," *American Economic Review* (December 1956) 46 (5): 835–56, and "The Micro-Foundations of Aggregate Demand and Supply," *Economic Journal* (September 1957) 67: 455–70.
3. The presentation in this chapter follows closely the views published by Michael L. Wachter, both in his own and jointly authored papers. See Michael L. Wachter, "The Changing Cyclical Responsiveness of Wage Inflation over the Postwar Period," *Brookings Papers on Economic Activity,* vol. 1 (1976), pp. 115–59; Michael L. Wachter and Susan M. Wachter, "The Fiscal Policy Dilemma: Cyclical Swings Dominated by Supply Side Constraints," in *The Economic Consequences of Slowing Population Growth,* T. Espenshade and W. Serow, eds. (New York: Academic Press, 1978), pp. 71–99; Richard A. Easterlin, Michael L. Wachter, and Susan M. Wach-

Notes

ter, "The Changing Impact of Population Swings on the American Economy," *Proceedings of the American Philosophical Society,* June 1978, vol. 122, no. 3, pp. 119–30; and Michael L. Wachter and Jeffrey M. Perloff, "A Production Function-Nonaccelerating Inflation Approach to Potential Output: Is Measured Potential Output Too High?" *Carnegie-Rochester Conference Series on Public Policy* (a supplement to *The Journal of Monetary Economics*), vol. 10, 1979.

4. Wachter and Perloff, "A Production Function-Nonaccelerating Inflation Approach."

5. Data sources are Bureau of Labor Statistics, *Employment and Earnings, January 1979,* U.S. Department of Labor, vol. 26, no. 1 (Washington, D.C.: U.S. Government Printing Office, 1979), p. 156.

6. Wachter and Perloff, "A Production Function-Nonaccelerating Inflation Approach."

7. Michael L. Wachter, "Unemployment Policies to Reduce Inflation," Discussion Paper no. 44, Center for the Study of Organizational Innovation, University of Pennsylvania, presented before the Joint Economic Committee, 9 February 1979.

8. *The 1979 Joint Economic Report: Report of the Joint Economic Committee, Congress of the United States on the 1979 Economic Report of the President* (Washington, D.C.: U.S. Government Printing Office, 1979), report no. 96–44, Senate.

9. For further discussion, see Michael L. Wachter and Susan M. Wachter, "The Fiscal Policy Dilemma."

Chapter 8

1. See the reference to studies by Wachter, Anderson, and Welch in chap. 2, note 5.

2. Michael L. Wachter, "Intermediate Swings in Labor Force Participation," *Brookings Papers on Economic Activity* (October 1977) 2: 545–76.

3. Ronald D. Lee, "Fluctuations in U.S. Fertility, Age Structure and Income," report to the National Institute of Child Health and Human Development, 1977; and the two papers by Ronald Lee cited in chap. 2, note 5; and David Goldberg, "Future of American Fertility: Some Speculations," paper presented to the Population Association of America, Atlanta, Ga., 1978. See also Jean Bourgeois-Pichat, "La Baisse Actuelle de la Fecondité en Europe: s'Inscrit-elle dans le Modele de la Transition Demographique?" *Population* (1979) 2: 267–306; Michael L. Wachter, "A Time Series Fertility Equation: The Potential for a Baby Boom in the 1980s," *International Economic Review* (October 1975) 16: 609–24; Peter Lindert, "American Fertility Patterns Since the Civil War," in *Population Patterns in the Past,* Ronald D. Lee, ed. (New York: Academic Press, 1977); Martin O'Connell, "The Effect of Changing Age Distributions on Fertility and Suicide in Developed Countries," unpublished Ph.D. dissertation in demography, University of Pennsylvania, 1975.

4. See the paper by James Taylor cited in chap. 6, note 1.

5. Department of Health, Education, and Welfare, *Divorces by Mar-*

Notes

riage Cohort, Vital and Health Statistics, series 21, no. 34 (Washington, D.C.: U.S. Government Printing Office, 1979), p. 15.

6. Technically, figure 3.3 shows the relation between the young adult proportion and the total fertility rate, a somewhat different fertility measure than the crude birth rate. The picture would look quite similar, however, if the crude birth rate were used. See Richard A. Easterlin, "What Will 1984 Be Like? Socioeconomic Implications of Recent Twists in Age Structure," *Demography* (November 1978) 15 (4): 397–432, fig. 4.

7. For some work along these lines, see P. A. Samuelson, "An Economist's Non-Linear Model for Self-Generated Fertility Waves," *Population Studies* (July 1976) 30 (2): 243–47; Frank T. Denton and Byron G. Spencer, *Population and the Economy* (Lexington, Mass.: Lexington Books, 1975); Nathan Keyfitz, "Population Waves," in *Population Dynamics,* T. N. E. Greville, ed. (New York: Academic Press, 1972); and the papers by Ronald Lee, cited in note 3, this chapter.

8. See also Easterlin, "What Will 1984 Be Like?" figure 9. The data in the text are derived from appendix table 1.1, the first entry and each succeeding fourth entry. Because of the discontinuity in the series, the 1955–60 value used to compute the change for the 1940–60 interval was raised by 0.4 points.

9. French demographer Jean Bourgeois-Pichat has recently suggested, on the basis of reasoning similar to mine, that self-generating fertility fluctuations are likely to be common in what demographers call the "final stage of the demographic transition." See Bourgeois-Pichat, "La Baisse Actuelle."

10. See for example, the citations of demographer Charles F. Westoff in chap. 3, note 1.

11. See Simon Kuznets, "Long Swings in the Growth of Population and in Related Economic Variables," *Proceedings of the American Philosophical Society* (1958) 102: 25–52; and Simon Kuznets, *Capital and the American Economy: Its Formation and Financing* (Princeton, N.J.: Princeton University Press, 1961); Moses Abramovitz, "The Nature and Significance of Kuznets Cycles," *Economic Development and Cultural Change* (April 1961) 9 (3): 225–48; Richard A. Easterlin, *Population, Labor Force, and Long Swings in Economic Growth: The American Experience* (New York: National Bureau of Economic Research, 1968).

12. For a full exposition, see Easterlin, ibid., chaps. 2 and 3.

13. Moses Abramovitz, "The Passing of the Kuznets Cycle," *Economica* (1968) 35: 349–67.

14. See Easterlin, "What Will 1984 Be Like?" Chart 1.

Chapter 9

1. Lance E. Davis et al., *American Economic Growth: An Economist's History of the United States* (New York: Harper & Row, 1972), p. 84.

2. On Sweden, see Shirley Weitz, *Sex Roles: Biological, Psychological and Social Foundations* (New York: Oxford University Press, 1977), pp. 228–29. See also Helen Mayer Hacker, "Gender Roles from a Cross-

Notes

Cultural Perspective," in Lucille Duberman, ed., *Gender and Sex in Society* (New York: Praeger, 1975), pp. 205–8. On the Soviet Union, see Weitz, ibid., pp. 207–11; Hacker, ibid., pp. 200–5; and Hedrick Smith, *The Russians* (New York: Ballantine, 1976), p. 179. See also Judith Blake, "The Changing Status of Women in Developed Countries," *Scientific American* (September 1974) 231 (3).

3. Charles F. Westoff, "Marriage and Fertility in the Developed Countries," *Scientific American* (December 1978) 239 (6): 52–53.

4. Several of the issues addressed here are raised in a recent paper by Oppenheimer from which I have benefited. See Valerie Kincade Oppenheimer, "Structural Sources of Economic Pressure for Wives to Work: An Analytical Framework," *Journal of Family History* (Summer 1979), pp. 177–97. See also "Life Cycle Squeeze: The Interaction of Men's Occupational and Family Life Cycles," *Demography* (May 1974) 11 (2): 227–45.

5. Michael L. Wachter, "The Labor Market and Immigration: The Outlook for the 1980's," prepared for the Interagency Task Force on Immigration, discussion paper no. 40, February 1979.

6. See Richard B. Freeman, *The Overeducated American* (New York: Academic Press, 1976), and Welch paper cited in chap. 2, note 5.

7. David Goldberg and A. Anderson, "Projections of Population and College Enrollments in Michigan, 1970–2000," paper presented to the Governor's Commission on Higher Education, Lansing, Mich., 1974.

8. See chapter 3, note 23.

9. Richard A. Easterlin, "Relative Economic Status and the American Fertility Swing," in Eleanor Bernert Sheldon, ed., *Family Economic Behavior: Problems and Prospects* (Philadelphia: J. B. Lippincott, 1972), p. 209.

10. George S. Masnick, Barbara Wiget, John R. Pitkin, and Dowell Meyers, "A Life Course Perspective on the Downturn in U.S. Fertility," Center for Population Studies, Harvard University, Cambridge, Mass., working paper no. 106, October 1978.

11. Anthony Downs, "Public Policy and the Rising Cost of Housing," Brookings General Series Reprint no. 344 (Washington, D.C.: 1978).

12. As quoted in Frederick Lewis Allen, *The Big Change* (New York: Harper, 1952).

13. Moses Abramovitz, "The Passing of the Kuznets Cycle," *Economica* (1968) 35: 367.

14. See, for example, James S. Duesenberry, *Income, Saving, and the Theory of Consumer Behavior* (Cambridge: Harvard University Press, 1966); Franco Modigliani, "Fluctuations in the Saving-Income Ratio: A Problem in Economic Forecasting," in Conference on Research in Income and Wealth, *Studies in Income and Wealth*, XI (New York: National Bureau of Economic Research, 1961), pp. 371–443; Emile Durkheim, *Suicide, A Study in Sociology* (New York: Free Press, 1951); Robert K. Merton, *Social Theory and Social Structure* 1968 enlarged ed. (New York, The Free Press, 1968); and S. A. Stouffer et al., *The American Soldier: Adjustment During Wartime Life,* vol. 1 (Princeton, N.J.: Princeton University Press, 1949); Leon Festinger, *A Theory of Cognitive Dissonance* (Evanston, Ill.: Row, Peterson, 1957).

15. Martin O'Connell, "The Effect of Changing Age Distribution on

Notes

Fertility: An International Comparison," in Julian Simon, ed., *Research in Population Economics*, vol. 1 (Urbana: University of Illinois Press, 1978); F. L. Sweetser and P. Peipponen, "Postwar Fertility Trends and Their Consequences in Finland and the U.S.," *Journal of Social History* (Winter 1967) 1: 108–18; Richard A. Easterlin and Gretchen A. Condran, "A Note on the Recent Fertility Swing in Australia, Canada, England and Wales, and the United States," in Hamish Richards, ed., *Population, Factor Movements, and Economic Development: Studies Presented to Brinley Thomas* (Cardiff, Great Britain: University of Wales Press, 1976); Helge Brunborg and G. S. Lettenstrøm, "Fertility Trends in Norway since 1965," paper presented to the Nordic Demographic Symposium, Rungsted, Denmark, 10–12 June 1976; H. Leridon, "Fecondité et Structures Demographiques: Une Hypothese sur l'Evolution de la Fecondité depuis 1940," *Population* (Mars/Avril 1978) 33: 441–47; see also Bourgeois-Pichat reference in chap. 8, note 3.

16. See Bourgeois-Pichat reference in chap. 8, note 3.

17. For an interesting illustration of this point, see the table summarizing policy constraints on layoffs in seven countries in Bennett Harrison, "Why the U.S. Unemployment Rate Is So High," *Challenge* (May–June 1978) 21 (2): 46.

18. Ronald R. Rindfuss and James A. Sweet, *Post-War Fertility Trends and Differentials in the United States* (New York: Academic Press, 1977).

19. Another educational topic where a numbers effect of a somewhat different nature may operate is SAT scores. See R. B. Zajonc, "Family Configuration and Intelligence," *Science* (1976) 192: 227–36.

20. For some formal analysis of this problem, see Alan Whitman, "Tenure Waves," *Journal of the Franklin Institute* (March 1975) 299 (3): 215–20; and Louis Henry, "Pyramides, Statuts et Carrières, I. Avancement a l'Ancienneté–Selection," *Population* (1971) 26 (3): 463–86, "II. Avancement au choix," *Population* (1972) 27 (4–5): 599–636, "III. Corps de petit effectif," *Population* (1976) 31 (4–5): 839–55. An excellent article that came to my attention after completing this book, relevant here and to cohort size theory generally, is Joan Waring, "Social Replenishment and Social Change," *American Behavioral Scientist* (November/December) 19 (2): 237–56.

21. Note the mention by Gertrude Bancroft of the breakdown more generally of traditional attitudes in this period in chap. 4, note 12.

INDEX

Abortion, 55, 90
Abramovitz, Moses, *x,* 157, 193, 194
Accidents, 106–7, 147, 190
Affirmative action, 150, 153
Age: and childbearing, 52–55; and criminal activity, 101–4; and economic welfare, 15–34; and hiring practices, 162; and labor market participation by women, 63, 66–73, 75; of retirement, 151; and unemployment rates, 114–27. *See also* Age-composition effect; Age-specific effect; Teenagers
Age-composition effect: defined, 6, 55; and indicators of social breakdown, 110–11; and unemployment rates, 117–18, 120
Age-specific effect, 6–14, 55; and unemployment rates, 118–20. *See also* Generation size
Aggregate demand, 113–27
Aging, 183
Airlines, 162
Alcohol abuse, 106–7, 147
Alienation, 4, 97, 108–10, 147; future cycles of, 134
Allen, Frederick Lewis, 194
Allen, Sheila, 185
Almond, Gabriel, 108, 191
Anderson, A., 194
Anderson, Joseph M., 136, 182, 192
Anomie, 160. *See also* Alienation
Arker, Sara, 186
Arteriosclerotic heart disease, 107

Asthma, 107
Atomic bombings, 156
Australia, 160

Baby boom. *See* Generation size
Bachman, Jerald G., 181
Bancroft, Gertrude, 72, 183, 186, 195
Barker, Diana, 185
Barnes, Carl B., 189
Barron, R. D., 185
Becker, Gary S., 187, 189
Belgium, 56, 160
Birth cohorts, 7, 183n
Birth control. *See* Contraception
Birth rates: and economic success, 19–29; and generation size, 4–14; patterns of change in, 7–8; and relative numbers at working age, 15–19; self-generating fluctuations in, 132–34, 137–44; and the supply of younger workers relative to older, 147–48. *See also* Childbearing; Fertility rates; Generation size; Marital fertility; Total fertility rate
Blake, Judith, 11, 181, 184, 185, 193
Bliven, Bruce, 156
Bourgeois-Pichat, Jean, 160, 192, 193, 195
Brier, Stephen S., 189

Brunborg, Helge, 195
Buckley, John E., 183
Butz, William P., 186

Canada, 160
Cancer, 107
Carithers, Martha, 188
Carter, Hugh, 187
Census Bureau studies, 91
Chafetz, Janet Salzman, 181, 185
Cherlin, Andrew, 187
Childbearing: attitudes toward, 76, 78; and the competitive disadvantage of women, 58–59; and contraception, 55–57; and divorce, 83, 87, 89, 93–94; economic determinants of, 39–52; and the education of women, 57–58; future cycles in, 134, 136, 138; and labor force participation by women, 60–78, 153–55; public policy to regulate, 163; and relative income, 83–89, 146–47; variations in, by age, 52–55; and World War II, 54. See also Birth rates; Fertility rates; Marital fertility; Total fertility rate
Child care centers, 154
Childlessness, 11, 93
Chilman, Catherine S., 187
Churches, 10
Cognitive dissonance, 76, 160
Cohen, Albert K., 189
Cohorts, 7, 183. See also Generation size
College enrollment, 152–53, 161–62
Commodity market problems, 127
Compulsory retirement, 151
Condom, 56
Condran, Gretchen A., x, 194
Conference Board, 64

Conservatism, 110
Construction industry, 152
Contraception, 61, 154; and fertility patterns, 38, 55–57; and labor force participation, 61, 78
Coombs, Jerry, x
Coombs, Lolagene C., 83, 89, 184, 188
Coyne, Lolafaye, 188
Credit, 114–15
Crime: future cycles in, 134, 157; and generation size, 4–6, 97, 101–6, 110–11, 147. See also Homicide
Crimmins, Eileen M., x
Crimmins-Gardner, Eileen, 184
Crouch, Robert, 189
Cutright, Phillips, 187
Czajka, John L., 186

David, Paul A., 189
Davis, James A., x
Davis, Lance E., 193
Demand manipulation, 113–27
Denton, Frank T., 193
Denmark, 160
Depressions, 32, 34; technology as a source of, 145–46
Depression, the, 86
Deviant behavior, 136. See also Crime; Homicide; Suicide
Diseases, 107
Divorce: future cycles in, 134, 136, 157; and generation size, 4–5, 80–89, 93–94, 110–11, 147; and relative income, 83–89; and the traditional family, 79–80
Douche, 56
Downgrading, 151–52
Downs, Anthony, 194
Drinking, 106–7
Drug use, 106–7

Index

Dual labor market, 63
Duberman, Lucille, 193
Duesenberry, James S., 160, 194
Durkheim, Emile, 160, 194

Easterlin, Richard A., 181, 182, 183, 184, 185, 186, 189, 191, 193, 194
Econometrics, 44, 189
Economic welfare: as determinant of marriage and childbearing patterns, 39–52; in the next two decades, 134–44, 157; and stagflation effects, 112–28; of young adults, 15–34. *See also* Relative income
Education: and employment, 152–53; and public policy, 163; of women, 65–66, 161–62. *See also* College enrollment; Educational institutions; Schools
Educational institutions, 25, 30–31. *See also* Education; Schools
Elder, Glen H., Jr., 184
Employment, 15–34. *See also* Labor market; Unemployment
Employment Act of 1946, 33, 113, 143, 146
Energy crises, 113, 127, 155
Entry-level jobs, 25–26
Environmental regulations, 127
Escalator clauses, 126
Espenshade, T., 191
Ewer, Phyllis, 184
Extramarital intercourse, 80
Eyer, Joseph, 188–89, 190, 191

Families: effects of generation size on, 13–14, 30; impact of divorce

and illegitimacy on, 79–94; impact of relative income on, 146–47; and patterns of marriage and childbearing, 37–38; persistence of traditional, 149–50; sex typing in, 9–14; size preferences for, 52, 154–55. *See also* Childbearing; Divorce; Marriage
Federal Center for Disease Control, 107
Federal youth job corps, 127
Feinberg, Stephen E., 189
Felson, Marcus, 189
Fertility rates: economic determinants of, 39–52; and homicide and suicide, 105–6; and labor force participation of women, 60–78; among minority groups, 161; in other nations, 160–61; self-generating fluctuations in, 193. *See also* Birth rates; Childbearing; Generation size; Marital fertility; Total fertility rate
Fertility regulation, 61. *See also* Contraception
Festinger, Leon, 160, 194
Finland, 160
Fox, James Alan, 189
France, 56, 160
Freedman, Deborah, 181
Freeman, Richard B., 182, 183, 194
Free market economic system, 161

Gallup poll, 182
Generation size: and crime, 4–6, 97, 101–6, 110–11, 147; defined, 4, 7–8; and divorce, 4–5, 80–89, 93–94, 110–11, 147; and the economic determinants of marriage and childbearing, 39–52; effect of, before World War II, 32–34; effect of, on young adults, 15–

Generation size *(continued)*
29; effect of, over the life cycle,
29–32; future cycles of, 131–44,
153–60; and illegitimacy, 89–94;
and the impact of stagflation,
128; institutional responses to,
161–62; and mental stress, 97–
101 *(see also* Mental stress);
and personal welfare, 3–14, 145–
48; and public policy, 163; and
sex typing, 9–14; and suicide,
97, 104–6. *See also* Birth rates;
Fertility rates
Germany, Federal Republic of, 160
Gibbs, Jack P., 189
Ginsberg, Ralph B., *x*
Glick, Paul C., 187
Goldberg, David, *x,* 136, 192, 194
Goode, William J., 187
Gordon, H. Scott, 31, 183
Government policies, 146; and generation size effects, 32–34; and
Kuznets cycles, 143–44; to ma-
nipulate aggregate demand, 113–
28; restraining layoffs, 195
Greville, T. N. E., 193

Hacker, Helen Mayer, 193
Hamermesh, Daniel S., 189
Harrison, Bennett, 195
Harter, Carl L., 182
Harvests, 113
Hauser, Robert M., x
Headaches, 97
Health problems, 97, 107
Health regulations, 127
Henry, Andrew F., 189
Henry, Louis, 195
Herzog, A. Regula, 181
Higgs, Robert, *x*
Hiring practices, 26–27, 162–63

Hiroshima, 156
Homeownership, 155
Homicide, 102–6, 136, 190
House, James S., 191
Housing, 142, 156
Hypertension, 107

Illegal immigration, 34, 151, 183
Illegitimacy, 188; future cycles in,
134, 136; and generation size,
4–5, 89–94, 110–11, 147; and
the traditional family, 79–80
Immigration, 5, 146; and generation size effect, 32; illegal, 34,
151, 183; and population growth
patterns, 142–44; unrestricted,
33
Industrial Revolution, 145
Inflation, 112–16, 128; and the
effect of age structure on unemployment, 116–27; and life decisions, 155–56. *See also* Stagflation
International Business Machines,
162
Intrauterine device (IUD), 55

Japan, 161
Job advancement, 25–26, 162–63
Job corps, 127
Job training programs, 163
Johnson, Lloyd D., 181
Johnston, Denis F., 17, 182

Kahne, Hilda, 185

Index

Kantner, John F., 91, 188
Keyfitz, Nathan, 193
Keynesian revolution, 112
Klebba, A. Joan, 190
Klein, Deborah Pisetzner, 185
Klein, Lawrence R., 191
Kohen, Andrew I., 185
Kuhn, James W., 110, 191
Kuznets cycle, 142–44, 148, 157
Kuznets, Simon, x, 140, 193

Labor market: downgrading in, 151–52; effect of age structure on, 114–27; effect of relative young worker scarcity on, 19–29; factors affecting participation of women in, 27–29, 60–78; and future generation size cycles, 134–36, 140–44; impact of illegal immigration on, 34, 151; new factors likely to influence, 150–55; older women in, 70–73; supply and demand effects on, 26, 113–27; before World War II, 32–34; younger women in, 66–70. See also Occupational distribution; Unemployment
Land, Kenneth C., x, 189
Landes, Elisabeth M., 187
Layoffs, 195n
Lebergott, Stanley, x, 183
Lee, Ronald D., x, 136, 182, 192, 193
Leridon, H., 195
Lettenstrøm, G. S., 195
Levis, 41
Lindert, Peter H., x, 192
Lipset, Seymour Martin, 191
Lynton, Edith F., 186
Lyster, William R., 105, 190

McDonald, John, 84–89, 188
Mandle, Joan D., 181, 186
Marital fertility, 48; current patterns in, 37–38. See also Birth rates; Childbearing; Fertility rates; Total fertility rate
Marriage: current patterns in, 37–38; economic determinants of, 39–52; effect of children on, 83; employment practices regarding, 162; and female college enrollment, 161–62; future cycles in, 134, 138; and generation size, 4; and relative income, 90, 146–47, 187. See also Divorce; Marital fertility
Masnick, George S., 194
Mason, Karen Oppenheim, x, 181, 186
Mason, William M., x, 191
Material aspirations, 40–42
Maternity leaves, 154
Mead, Margaret, 38, 184
Media, 10
Men: and family sex typing, 9–14; suicide among young, 104
Mental stress: and drug use, 106; and generation size, 4–5, 97–101, 147; indicated by crime and suicide, 104–6; in the next two decades, 135, 137; and the rate of mortality, 107
Merton, Robert K., 160, 187, 189, 194
Meyers, Dowell, 194
Michael, Robert T., 187
Migrant agricultural workers, 34
Migration, 16
Millar, J. Donald, 107
Minimum wage laws, 25, 27
Minority groups, 153, 161. See also Race
Modigliani, Franco, 160, 194
Moonlighting, 68–69, 70n

Index

Moore, Maurice J., 91–92, 188
Morning-after pill, 61
Mortality, 16; and generation size, 106–7
Motor vehicle accidents, 107, 147, 190
Murder, 102–6, 136, 190

Nagasaki, 156
National Institute of Alcohol Abuse and Alcoholism, 106
Nervousness, 97. *See also* Mental stress
New York Times, The, 20, 23
Nisbet, Robert, 187, 189
Norris, G. M., 185
Norton, Arthur J., 187
Norway, 160

Occupational distribution: sex polarization in, 185; of women, 29, 58–59, 62–65, 183
Occupational orientation, 10–13
Occupational Outlook Handbook, 59, 75
O'Connell, Martin, *x,* 91–92, 105–6, 188, 189, 192, 194
Office of Management and Budget, 17
Oppenheimer, Valerie K., *x,* 183, 194
Oral birth control pills, 55–57

Parke, Robert, Jr., *x,* 181, 187
Part-time work, 154

Pathologies, 107
Peer group influences: and sex role images, 10; and teenage rebelliousness, 108
Peipponen, P., 194
Perloff, Jeffrey, 124, 191–92
Personal welfare, 3–14, 145–48
Physiological problems, 107
Pill, the, 55–57
Pitkin, John R., 194
Pogrebin, Letty Cottin, 181
Political alienation. *See* Alienation
Pollak, Robert A., *x*
Porterfield, Austin L., 190
Premarital conceptions, 90–93, 188
Premarital intercourse, 80
President's Commission on Population Growth and the American Future, 37
Preston, Samuel H., 84–89, 188
Promotion, 25–26, 162–63
Psychological stress. *See* Mental stress
Public policy, 163. *See also* Government policies

Race: and fertility rates, 161; and homicide, 104
Rainwater, Lee, 42, 184
Rand Corporation, 74
Reder, Melvin W., 189
Reiss, Albert J., Jr., *x*
Relative deprivation, 160
Relative income, 160; as a determinant of childbearing, 41–52; and illegitimacy, 90–93; impact of, on families, 146–47; and legitimation of premarital conceptions, 92–93; and marital stability, 187*n*; and mental stress indicators, 101–6, 110–11; and

Index

participation of women in the labor force, 60–62, 68–70, 73–74, 77–78; and role fulfillment in marriages, 80–89; and self-generating birth rate cycles, 137–39. *See also* Economic welfare
Religion, 10, 83, 159
Retirement, 151
Rhythm method, 55–56
Richards, Hamish, 195
Ridley, Jeanne Clare, 56, 184, 185
Riley, Matilda White, 183
Rindfuss, Ronald R., 184, 195
Robinson, John P., 189
Role fulfillment, 80–89
Role models, 9–11
Rossi, Alice S., 9, 181
Rumania, 31
Ryder, Norman B., 56–57, 154, 184, 185

Safety regulations, 127
Samuelson, P. A., 193
SAT scores, 31, 195
Scanzoni, John, 187
Schapiro, Morton Owen, *xi*
Schools, 30–31; sex typing in, 10. *See also* College enrollment; Education
Self-employment, 152
Self-image, 98
Serow, W., 191
Sex roles, 9–14, 149–50
Sexual intercourse, 90–93; extra-marital, 80
Sexual revolution, 80
Shaeffer, Ruth Gilbert, 186
Shaver, Phillip R., 189
Sheldon, Eleanor Bernert, 185, 194
Short, James F., Jr., 189

Simon, Julian L., 189, 194
Skolnick, Arlene, 187
Sibling rivalry, 30
Smith, Hedrick, 193
Social deterioration, 97, 110–11, 147–48. *See also* Alienation; Crime; Divorce; Homicide; Suicide
Social Indicators Project, 17
Socialization: and the formation of material aspirations, 40–44; and sex typing, 9–14, 149–50
Social Security system, 151
Soss, N., 189
Soviet Union, 149, 161
Spencer, Byron G., 193
Stagflation, 112–28; future patterns of, 134, 157; and generation size, 4–5
Stanford University, 108
Stetson, Dorothy M., 187
Stewardesses, 162
Stouffer, S. A., 160, 194
Student activism, 108
Suicide, 147, 189, 190; future cycles of, 157; and generation size, 4–5, 97, 104–6
Supply conditions, 113–27
Survey Research Center (University of Michigan), 99
Sweden, 149, 160
Sweet, James A., 184, 195
Sweetser, F. L., 194

Taffel, Selma, 184
Tax-relief measures, 114
Taylor, James B., 136, 188, 192
Technology, 145
Teenagers: death rates for, 107; illegitimacy among, 90–93, 188;

Teenagers *(continued)*
rebelliousness among, 108; and recent homicide rates, 136; school enrollment of female, 162
Television, 10
Tenure, 162
Thornton, Arland, 181
Total fertility rate, 192; defined, 48*n*; and oral contraceptive use, 56; and relative income, 48–50. *See also* Birth rates; Childbearing; Fertility rates; Marital fertility
Traffic fatalities, 190. *See also* Motor vehicle accidents
Tucker, Graham S. L., *x*
Turnover rates, 185*n*

Ulcers, 107
Unemployment: effect of age structure on, 114–27; future cycles of, 135–36; and generation size, 4–6, 20, 23, 25, 30; government policy effects on, 33–34, 113–28; and inflation, 112–28; noninflationary rate of, 123–26; among women, 27–28, 120. *See also* Stagflation
Unemployment compensation, 25, 27
Unions, 27, 113–14, 152
United Kingdom, 160
University of Michigan, 83, 99, 136
University of Pennsylvania, 124
Urban services, 142

Veroff, Joseph, 99, 189

Vietnam War, 108
Vigderhous, Gideon, 189
Violence, 102–7. *See also* Murder

Wachter, Michael L., x, 124, 136, 182, 183, 186, 191–92, 194
Wachter, Susan M., 183, 191, 192
Wage contracts, 113–14, 127
Waldron, Ingrid, 188–89, 190, 191
Waller, Willard, 81
War, 108, 156. *See also* World War II
Ward, Michael P., 186
Waring, Joan, 195
Watergate, 108
Weintraub, Sidney, 191
Weiss, Noel S., 190
Weitz, Shirley, 193
Welch, Finis, 136, 182, 183, 192
Wellford, Charles F., 102, 190
Westoff, Charles F., 37–38, 54, 56, 181, 183–84, 185, 186, 187, 193, 194
Whelpton, Pascal K., 184
Whitman, Alan, 195
Wiget, Barbara, 194
Wilson, James Q., 189
Withdrawal, 56
Wolpin, Kenneth I., 189
Women: changing status of, 37–38, 149–50; education of, 57–58, 65–66; effect of family sex typing on, 9–14; effect on, of unemployment among men, 114, 119–21; future work patterns of, 134, 136, 153–55, 158; labor market participation of, 4, 27–29, 59–78; marriage and college enrollment among, 161–62; occupational distribution of, 29, 58–59, 62–65, 183; suicide among,

Index

104; turnover rates among, 185; unemployment among, 27–28, 120

Women's movement, 76–78

World War II: and contraception education, 56; divorce during, 86; effect of, on labor market participation of women, 71–73; fertility rates following, 54

Wright, Gerald C., Jr., 187

Youth job corps, 127

Zajonc, Robert, 186, 195

Zelnik, Melvin, 91, 188

Zero population growth ideology, 76–78

Zumeta, Zena, 83–84, 89, 184, 188